159
Advances in Polymer Science

Editorial Board:
A. Abe · A.-C. Albertsson · H.-J. Cantow
K. Dušek · S. Edwards · H. Höcker
J. F. Joanny · H.-H. Kausch · K.-S. Lee
L. Monnerie · S. I. Stupp · U. W. Suter
G. Wegner · R. J. Young

Statistical, Gradient, Block and Graft Copolymers by Controlled/Living Radical Polymerizations

By Kelly A. Davis, Krzysztof Matyjaszewski

 Springer

Authors

Prof. Krzysztof Matyjaszewski
Dept. of Chemistry
Carnegie Mellon University
4400 Fifth Avenue
Pittsburgh, PA 15213
USA
E-mail: km3b@andrew.cmu.edu

Dr. Kelly A. Davis
Howard Hughes Medical Institute
University of Colorado-Boulder
Campus Box 424
Boulder, CO 80309
USA
E-mail: Kelly.Davis@Colorado.edu

This series presents critical reviews of the present and future trends in polymer and biopolymer science including chemistry, physical chemistry, physics and materials science. It is addressed to all scientists at universities and in industry who wish to keep abreast of advances in the topics covered.

As a rule, contributions are specially commissioned. The editors and publishers will, however, always be pleased to receive suggestions and supplementary information. Papers are accepted for „Advances in Polymer Science" in English.

In references Advances in Polymer Science is abbreviated Adv Polym Sci and is cited as a journal.

Springer APS home page: http://link.springer.de/series/aps/ or
http://link.springer-ny.com/series/aps/
Springer-Verlag home page: http://www.springer.de

ISSN 0065-3195
ISBN 3-540-43244-2
Springer-Verlag Berlin Heidelberg New York

Library of Congress Catalog Card Number 61642

This work is subject to copyright. All rights are reserved, whether the whole or part of the material is concerned, specifically the rights of translation, reprinting, re-use of illustrations, recitation, broadcasting, reproduction on microfilms or in other ways, and storage in data banks. Duplication of this publication or parts thereof is only permitted under the provisions of the German Copyright Law of September 9, 1965, in its current version, and permission for use must always be obtained from Springer-Verlag. Violations are liable for prosecution under the German Copyright Law.

Springer-Verlag Berlin Heidelberg New York
a member of BertelsmannSpringer Science+Business Media GmbH
http://www.springer.de

© Springer-Verlag Berlin Heidelberg 2002
Printed in Germany

The use of registered names, trademarks, etc. in this publication does not imply, even in the absence of a specific statement, that such names are exempt from the relevant protective laws and regulations and therefore free for general use.

Typesetting: Data conversion by MEDIO, Berlin
Cover: MEDIO, Berlin
Printed on acid-free paper   SPIN: 10856712   02/3020wei - 5 4 3 2 1 0

Editorial Board

Prof. Akihiro Abe
Department of Industrial Chemistry
Tokyo Institute of Polytechnics
1583 Iiyama, Atsugi-shi 243-02, Japan
E-mail: aabe@chem.t-kougei.ac.jp

Prof. Ann-Christine Albertsson
Department of Polymer Technology
The Royal Institute of Technolgy
S-10044 Stockholm, Sweden
E-mail: aila@polymer.kth.se

Prof. Hans-Joachim Cantow
Freiburger Materialforschungszentrum
Stefan Meier-Str. 21
79104 Freiburg i. Br., Germany
E-mail: cantow@fmf.uni-freiburg.de

Prof. Karel Dušek
Institute of Macromolecular Chemistry, Czech
Academy of Sciences of the Czech Republic
Heyrovský Sq. 2
16206 Prague 6, Czech Republic
E-mail: dusek@imc.cas.cz

Prof. Sam Edwards
Department of Physics
Cavendish Laboratory
University of Cambridge
Madingley Road
Cambridge CB3 OHE, UK
E-mail: sfe11@phy.cam.ac.uk

Prof. Hartwig Höcker
Lehrstuhl für Textilchemie
und Makromolekulare Chemie
RWTH Aachen
Veltmanplatz 8
52062 Aachen, Germany
E-mail: hoecker@dwi.rwth-aachen.de

Prof. Jean-François Joanny
Institute Charles Sadron
6, rue Boussingault
F-67083 Strasbourg Cedex, France
E-mail: joanny@europe.u-strasbg.fr

Prof. Hans-Henning Kausch
c/o IGC I, Lab. of Polyelectrolytes
and Biomacromolecules
EPFL-Ecublens
CH-1015 Lausanne, Switzerland
E-mail: kausch.cully@bluewin.ch

Prof. Kwang-Sup Lee
Department of Polymer Science & Engineering
Hannam University
133 Ojung-Dong
Teajon 300-791, Korea
E-mail: kslee@mail.hannam.ac.kr

Prof. Lucien Monnerie
École Supérieure de Physique et de Chimie
Industrielles
Laboratoire de Physico-Chimie
Structurale et Macromoléculaire
10, rue Vauquelin
75231 Paris Cedex 05, France
E-mail: lucien.monnerie@espci.fr

Prof. Samuel I. Stupp
Department of Measurement Materials Science
and Engineering
Northwestern University
2225 North Campus Drive
Evanston, IL 60208-3113, USA
E-mail: s-stupp@nwu.edu

Prof. Ulrich W. Suter
Department of Materials
Institute of Polymers
ETZ,CNB E92
CH-8092 Zürich, Switzerland
E-mail: suter@ifp.mat.ethz.ch

Prof. Gerhard Wegner
Max-Planck-Institut für Polymerforschung
Ackermannweg 10
Postfach 3148
55128 Mainz, Germany
E-mail: wegner@mpip-mainz.mpg.de

Prof. Robert J. Young
Manchester Materials Science Centre
University of Manchester and UMIST
Grosvenor Street
Manchester M1 7HS, UK
E-mail: robert.young@umist.ac.uk

Foreword

The design and the realisation of well-defined polymer architectures has become an important goal in macromolecular science. The prerequisite for achieving this goal is the availability of controlled polymerisation reactions. Living anionic polymerisation was the first reaction fulfilling these requirements. Cationic polymerisation only came into play when it was realised that it was possible to create an equilibrium between active and dormant species with the fraction of the dormant species being far superior to that of active ones.

A corresponding principle applies to controlled radical polymerisation performed in quite a number of modes such as nitroxide-mediated polymerisation (NMP), atom transfer radical polymerisation (ATRP), reversible addition fragmentation chain transfer (RAFT) or catalytic chain transfer (CCT) reactions. All of these variants of controlled radical polymerisation lead to well-defined architectures with the particular advantage that a much larger number of monomers are suitable and the reaction conditions are much less demanding than those of living ionic polymerisation reactions.

Although in controlled radical polymerisation, termination reactions cannot be excluded completely, they are limited in their extent and consequently the molecular weight is controlled, the polydispersity index is small and functionalities can be attached to the macromolecules. These features are indicative of the realisation of well-defined polymer architectures such as block copolymers, star-shaped and comb-shaped copolymers.

The present volume is particularly concerned with the use of the different modes of controlled radical polymerisation for the preparation of copolymers such as random copolymers, linear block copolymers, as well as graft copolymers and star-shaped copolymers. It also presents the combination of controlled radical polymerisation with non-controlled radical copolymerisation, cationic and anionic polymerisation, both of vinyl monomers and cyclic monomers, and ring-opening metathesis polymerisation.

The power of controlled radical polymerisation is demonstrated convincingly and the limitations of the synthetic approaches clearly indicated.

Last but not least the volume presents some potential applications for copolymers obtained by controlled radical polymerisation. It is expected that the first commercial products will appear on the market this year, giving convincing evidence for the importance of controlled radical polymerisation methods.

Aachen, March 2002   Hartwig Höcker

Advances in Polymer Science
Available Electronically

For all customers with a standing order for Advances in Polymer Science we offer the electronic form via LINK free of charge. Please contact your librarian who can receive a password for free access to the full articles. By registration at:

http://link.springer.de/series/aps/reg_form.htm

If you do not have a standing order you can nevertheless browse through the table of contents of the volumes and the abstracts of each article at:

http://link.springer.de/series/aps/
http://link.springer-ny.com/series/aps/

There you will find also information about the

- Editorial Board
- Aims and Scope
- Instructions for Authors
- Sample Contribution

Contents

1	Background	2
1.1	Copolymers	2
1.2	Free Radical Polymerization	3
1.3	Controlled/Living Radical Polymerization (CRP)	5
1.3.1	Stable Free Radical Polymerization and Nitroxide Mediated Polymerization (SFRP and NMP)	8
1.3.2	Atom Transfer Radical Polymerization (ATRP)	9
1.3.3	Degenerative Chain Transfer Including RAFT	10
1.4	Summary	11
2	Statistical Copolymers	14
2.1	SFRP/NMP	15
2.2	ATRP	19
2.3	Degenerative Transfer Processes	27
2.4	Comparison of Various CRP Methods Applied to Statistical Copolymers	27
3	Linear Block Copolymers	30
3.1	Linear Block Copolymers Prepared Exclusively by CRP Methods	30
3.1.1	SFRP/NMP	30
3.1.2	ATRP	44
3.1.3	Degenerative Transfer Processes	68
3.1.4	Comparison of CRP Methods for Block Copolymer Synthesis	70
3.2	Block Copolymers Prepared Through Transformation Techniques	72
3.2.1	CRP from Commercially Available Macroinitiators	72
3.2.2	Block Copolymers by Combination of Two Polymerization Techniques	79
3.2.3	Summary	103

4	**Other Chain Architectures**	107
4.1	Graft Copolymers	107
4.1.1	Grafting From	108
4.1.2	Grafting Through	117
4.1.3	Grafting Onto	126
4.1.4	Grafting from Surfaces	127
4.1.5	Summary	137
4.2	Star Polymers	138
4.3	Simultaneous/Dual Living Polymerizations	147
5	**Overall Summary**	153
5.1	General Overview	153
5.2	Critical Evaluation of CRP Methods for Materials Preparation	153
5.3	Potential Applications for Copolymers Made by CRP Methods	155
	References	157
	List of Abbreviations	166
	Author Index Volumes 101-159	171
	Subject Index	185

Statistical, Gradient, Block, and Graft Copolymers by Controlled/Living Radical Polymerizations

Kelly A. Davis[1] · Krzysztof Matyjaszewski[2]

[1] Howard Hughes Medical Institute, University of Colorado-Boulder, Campus Box 424, Boulder, CO 80309, USA
E-mail: Kelly.Davis@Colorado.edu
[2] Center for Macromolecular Engineering, Department of Chemistry, Carnegie Mellon University, 4400 Fifth Ave., Pittsburgh, PA 15213, USA
E-mail: km3b@andrew.cmu.edu

This review is focused on controlled/living radical polymerization methods for the preparation of various copolymers. A brief introduction to the subject of radical polymerization, and early attempts to control it, is followed by a detailed examination of the literature on controlled/living radical copolymerizations from the mid-1990s until 2001. The topics covered include statistical/gradient, block, graft, and star copolymers, and the polymerization methods used to produce them. These copolymers were prepared using three major controlled radical methods (either nitroxide mediated polymerization, atom transfer radical polymerization, or degenerative transfer) and a combination of polymerization techniques, including transformation chemistry or the simultaneous/dual living polymerizations, to achieve the desired chain architecture or topology. An evaluation of the current state of the field is also presented.

Keywords: Review, Copolymer, Controlled/living radical polymerization, Block, Graft, Gradient, Statistical

1
Background

1.1
Copolymers

Copolymers, for the purpose of this review, are defined as macromolecules that contain more than one type of monomer unit within the polymer chain. There are many types of copolymers that fall into this category, as illustrated in Fig. 1.

They include the chain topologies of statistical (also periodic), gradient, and segmented copolymers (blocks and grafts); however, when the chain architecture is varied to include comb, multi-arm stars and dendrimers, or even growth from functionalized surfaces, the possibilities for compositional modification of different copolymers are almost limitless. Historically, many copolymers have been utilized as stabilizers for polymer blends or for latexes, but because of ill-defined compositions and properties, the particular reason behind failure in any specific application was not readily obvious. With the advent of ionic living polymerization, well-defined polymers have become the norm rather than the exception. This allows a structure-property correlation to be developed based on composition, chain topology, or architecture, and provides substantial information regarding how small changes in any of these parameters affect the resulting properties.

Living ionic methods, however, have limitations as to the types of monomers that can be polymerized resulting from the incompatibility between the reactive centers and monomers. Radical polymerizations, on the other hand, do not really suffer from these drawbacks because a free radical is less discriminating re-

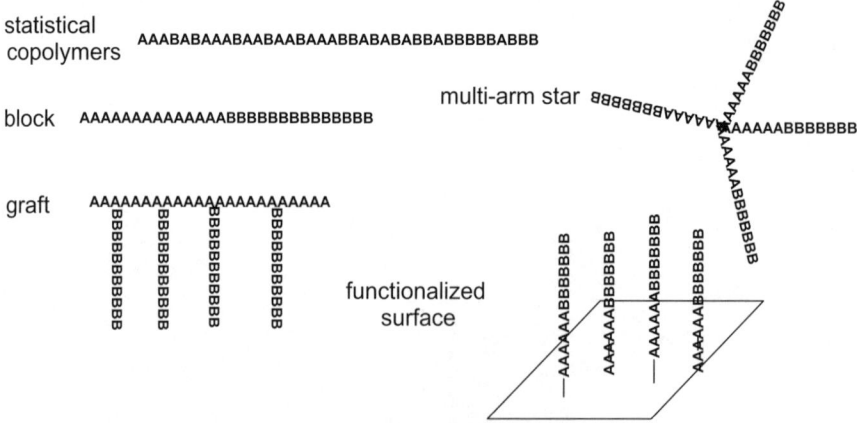

Fig. 1. Illustration of different types of segmented copolymers

garding the types of vinyl monomers with which it can react and is tolerant to many functionalities. This advantage allows for the preparation of statistical (we use this general term rather than the term random, which refers to a Bernoullian distribution) and segmented copolymers not possible with ionic methods, like various combinations of acrylate and methacrylate based monomers. The next section introduces the concepts behind radical polymerizations in general, followed by specifics about early attempts to control them, then by background information about newly developed controlled/living radical polymerization (CRP) methods. Subsequent sections will discuss in detail the preparation of segmented copolymers using CRP methodologies.

1.2
Free Radical Polymerization

Free radical polymerization is an integral part of polymer chemistry [1–4]. It has become a widely used industrial methodology because generation of a radical is easy, many monomers can be polymerized, and radical polymerizations are tolerant to the impurities that normally would terminate an ionic polymerization (moisture, protic solvents), making it an economically attractive alternative to the rigorous purification needed in ionic processes. The drawback of radical polymerizations, however, is that while it is easy to generate a reactive radical that can initiate polymerization, the polymerization itself is difficult to control. Unlike ionic species that repel one another, a radical will readily react with another radical, either through disproportionation or through a coupling reaction. The former will produce both a saturated and an unsaturated chain end, while the latter will produce a single dead chain (Scheme 1).

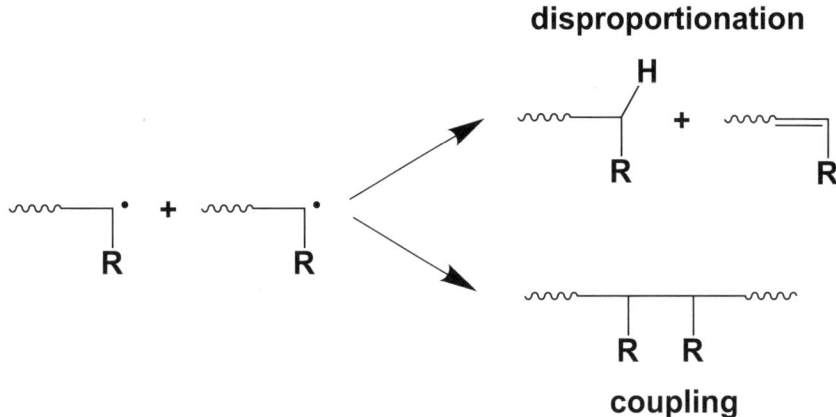

Scheme 1. Illustration of the modes of termination in radical polymerization

Termination reactions cannot be eliminated in radical polymerizations because termination reactions involve the same active radical species as propagation; therefore, eliminating the species that participates in termination would also result in no polymerization. Termination between active propagating species in cationic or anionic processes does not occur to the same extent because of electrostatic repulsions. Equation (1) represents the rate of polymerization, R_p, which is first order with respect to the concentration of monomer, M, and radicals, P*, while Eq. (2) defines the rate of termination, R_t, which is second order with respect to the concentration of radicals. To grow polymer chains with a degree of polymerization of 1000, the rate of propagation must be at least 1000 times faster than the rate of termination (which under steady state condition is equal to the rate of initiation). This requires a very low concentration of radicals to minimize the influence of termination. However, termination eventually prevails and all the polymer chains produced in a conventional free radical process will be "dead" chains. Therefore they cannot be used in further reactions unless they contain some functional unit from the initiator or a chain transfer agent.

$$R_p = k_p [M] [P^*] \tag{1}$$

$$R_t = k_t [P^*]^2 \tag{2}$$

Another major limitation in conventional radical polymerizations is that the molecular weight of the polymer chains is poorly controlled. Most free radical initiators have an initiator efficiency <1. There are several reasons for this, including the cage effect and primary radical termination [3]. As a consequence of using thermally activated initiators for the polymerizations, which have a long half lifetime at a given temperature, very few polymer chains are initiated at the onset of polymerization and initiation continues throughout the polymerization, resulting in a broad distribution of chain lengths and ill-defined polymers. Typical initiators include 2,2′-azobisisobutyronitrile (AIBN) which has a half lifetime of ~10 h at 65°C [3]. This means that only half the initiator will be consumed after ~10 h, leaving a significant portion that will continue to decompose and begin new chains. If the temperature is increased to enhance the decomposition rate, the termination rate will also increase because the termination rate is dependent on the concentration of radicals, leading to a significant decrease in the molecular weight.

There have been attempts to remedy this situation. Chain transfer agents can be used to limit the molecular weight of the polymer chains. These additives react with the growing polymer chain, limiting the degree of polymerization without affecting the polymerization rate. Transfer agents can introduce functionality to polymer chain ends that will allow for post-polymerization reactions.

The reactivity ratios of various monomer combinations in the free radical copolymerizations have been determined [5]. The reactivity ratio, r_1, is defined as

1 Background

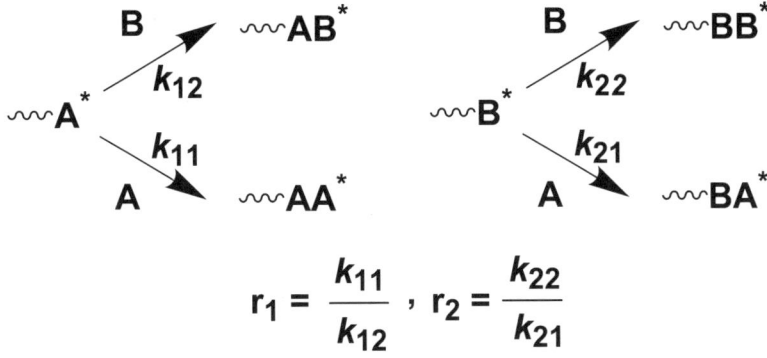

Scheme 2. Reactivity ratios

the ratio of the rate constant of homopropagation, k_{11}, to that of cross-propagation, k_{12} (Scheme 2). The reactivity ratios dictate the composition and microstructure of the polymer backbone and are specific to a given process, and may be different for each process, whether it is an anionic, cationic, or radical polymerization. Controlled/living radical polymerizations have been used to prepare a class of copolymers, referred to as tapered or gradient copolymers [6–8], where the instantaneous composition along the polymer backbone varies as a function of the monomer feed and the reactivity ratios for the given monomers. However, in conventional radical polymerizations, the slow continuous initiation results in copolymers where the composition varies among the chains as a function of the instantaneous monomer concentration in the reaction mixture. Obtaining full control over the free radical processes allows preparation of polymers that are not only well-defined compositionally along the chain, but among the chains as well.

1.3
Controlled/Living Radical Polymerization (CRP)

The concept of living polymerizations started in 1956 when Szwarc found that in the anionic polymerizations of styrene (St) the polymer chains grew until all the monomer was consumed [9], and that the chains continued growing when another batch of monomer was added. The addition of another monomer resulted in the formation of block copolymers. These polymerizations proceeded without termination or chain transfer occurring in the system. Prior to this work, the conditions used for the polymerizations had not been stringent enough to keep the active species alive and allow observation of this type of behavior. The polymer molecular weights were predictable based on the ratio of monomer to initiator and the polydispersities were low, indicating the polymerization was well controlled.

Later, other living systems were also achieved [10]. These include ring opening polymerization [11, 12] and carbocationic systems [13, 14]. In the carbocationic systems the high reactivity of the active species required that an equilibrium between the "active" species and a "dormant" species be formed, thus allowing control over the polymerization [13, 15, 16]. This approach was subsequently extended to controlled/living radical polymerizations (CRPs) [17–19]. Conceptually, if there is only a tiny amount (ppm) of chains that are active at any given point in time while the others are dormant, this lessens the overall effect of termination. Although termination cannot be avoided, at the same polymerization rate (i.e., the same radical concentration), the same number of chains terminate, but the percentage of dead chains relative to the total number of growing chains would be very small (<10%). This is because while the total number of chains in the conventional process equals the sum of dead and propagating chains, in CRP the total number is the sum of dead ($\sim 10^{-3}$ mol/l), propagating ($\sim 10^{-8}$ mol/l), and dormant ($\sim 10^{-2}$ mol/l) chains. The presence of dormant chains that are still functional provides a route to segmented copolymers without the need for additional transfer agents.

It must be stressed that in CRPs as well as in many other new "living" systems, termination cannot be fully suppressed and these systems should be distinguished carefully from true living polymerizations [20–22]. The imperfections in chain end functionalities and blocking efficiency may not be detrimental to the materials properties, but deviations from the ideal systems should be quantified [23, 24].

There were several early attempts to regulate free radical polymerizations [3, 25–28]. These methods utilized so-called iniferters, i.e., compounds which could serve as INItiators, transFER agents and TERminating agents. They could be activated photochemically [19, 29, 30] or thermally. In the latter case, bulky organic moieties based on diaryl or triarylmethyl derivatives were used [19, 31–34]. These types of systems have been studied for the formation of segmented copolymers, mostly for block copolymers. These techniques, while useful, did not offer the desired level of control over the polymerizations due to poor molecular weight control, high polydispersities, and low blocking efficiency. The main disadvantages of these systems comprise slow initiation, slow exchange, direct reaction of "counterradicals" with monomers, and their thermal decomposition.

New methods were developed in the mid-1990s based on the idea of establishing an equilibrium between the active and dormant species [35, 36]. Three approaches were the most successful:
1. Control via a reversible homolytic cleavage of a weak covalent bond leading to a propagating radical and a stable free radical. The latter should only react with the propagating radical and can be a nitroxide [37, 38], an N-based radical [39], or an organometallic species [40, 41]. They are generally called sta-

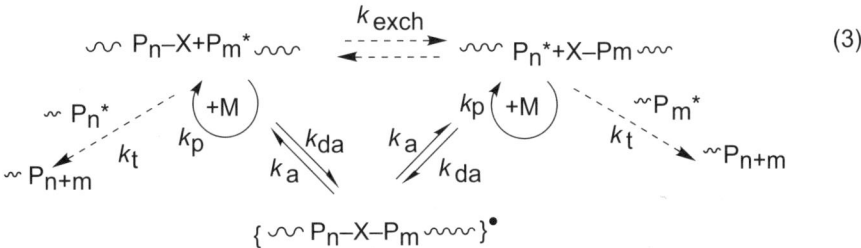

Fig. 2. The mechanisms of (1) stable free radical polymerizations, (2) reversible redox polymerizations (i.e., ATRP), and (3) degenerative chain transfer

ble free radical polymerizations (SFRP) or nitroxide mediated processes (NMP) (1 in Fig. 2)
2. Control via a reversible redox reaction between alkyl halides and transition metal complexes, i.e., atom transfer radical polymerization (ATRP) [42–49] (2 in Fig. 2)
3. Degenerative chain transfer with alkyl iodides [50, 51] or dithioesters (RAFT or MADIX) [52–55] (3 in Fig. 2)

The mechanism invoked in these CRP processes to extend the lifetime of growing radicals from ~1 s to a few hours utilizes a dynamic equilibration between dormant and active sites with a rapid exchange between the two states. Unlike conventional radical processes, CRP requires the use of a persistent radical (deactivator) species, or highly active transfer agents to react with propagating radicals. The persistent radicals/transfer agents react with radicals (deactivation or transfer reactions with rate constant, k_d) to form the dormant species. Conversely, propagating radicals are generated from the dormant species by an activation reaction (with rate constant, k_a).

While these three systems possess different components, the general similarities between the CRP processes are in the use of initiators, radical mediators (i.e., persistent radicals or transfer agents), and in some cases, catalysts (Fig. 2). It is important to note that while SFRP/NMP and ATRP are subject to the Persistent Radical Effect (PRE) [56] the degenerative processes, such as RAFT, do not conform to the PRE model due to the transfer dominated nature of the reaction.

1.3.1
Stable Free Radical Polymerization and Nitroxide Mediated Polymerization (SFRP and NMP)

In 1993, Georges et al. reported on the controlled radical polymerization of St initiated by benzoyl peroxide and mediated by 2,2,6,6-tetramethyl-1-piperidinyloxyl (TEMPO), a stable nitroxide radical [38]. TEMPO was able to bond reversibly to the polystyryl chain end and provide polystyrenes with predetermined molecular weights and low polydispersities. Nitroxides used earlier to control radical polymerizations were less successful [37, 57]. Scheme 3 illustrates the mechanism of the St polymerization, using a generalized structure of radical initiator I-I, and details the structure of TEMPO. Although several types of nitrox-

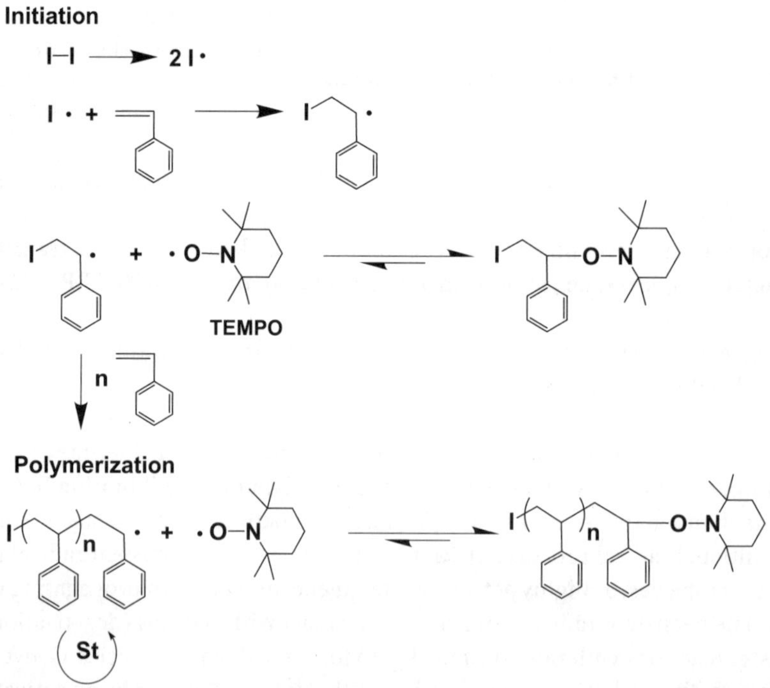

Scheme 3. Mechanism of polymerization of styrene using TEMPO-mediated CRP

ides had been synthesized [58, 59] and their ability to trap radicals reversibly was known [60], this was the first open literature report on using TEMPO to moderate a polymerization successfully [37]. Unfortunately, TEMPO can only be used for the polymerization of styrene-based monomers at relatively high temperatures (>120°C). With most other monomers, the bond formed is too stable and TEMPO acts as an inhibitor in the polymerization, preventing chain growth. With methacrylates, β-hydrogen abstraction results in a stable hydroxylamine and unsaturated chain ends. The radicals generated from the thermal self-initiation of St also help to control the rate of polymerization by reacting with any excess TEMPO that forms due to termination reactions, i.e., operates via persistent radical effect (PRE) [56].

Since TEMPO is only a regulator, not an initiator, radicals must be generated from another source; the required amount of TEMPO depends on the initiator efficiency. Application of alkoxyamines (i.e., unimolecular initiators) allows for stoichiometric amounts of the initiating and mediating species to be incorporated and enables the use of multifunctional initiators, growing chains in several directions [61]. Numerous advances have been made in both the synthesis of different types of unimolecular initiators (alkoxyamines) that can be used not only for the polymerization of St-based monomers, but other monomers as well [62–69]. Most recently, the use of more reactive alkoxyamines and less reactive nitroxides has expanded the range of polymerizable monomers to acrylates, dienes, and acrylamides [70–73]. An important issue is the stability of nitroxides and other stable radicals. Apparently, slow self-destruction of the PRE helps control the polymerization [39]. Specific details about use of stable free radicals for the synthesis of copolymers can be found in later sections.

1.3.2
Atom Transfer Radical Polymerization (ATRP)

The concept of using transition metal complexes to mediate radical polymerizations developed out of atom transfer radical addition reactions (ATRA), also referred to as the Kharasch reaction, which originally used light to generate a radical [74]. Later, transition metal complexes were used to promote halogen addition to alkenes through a redox process [75, 76]. As shown in Scheme 4, a lower oxidation state metal abstracts a halogen from an activated alkyl halide, which can then add across the double bond of an alkene. The newly formed radical re-abstracts the halogen from the higher oxidation state metal to form an alkene-alkyl halide adduct and regenerate the lower oxidation state metal. In efficient ATRA, trapping of the product radical should be faster than the subsequent propagation step and reactivation of the adduct should be very slow, maximizing the yield of the targeted product. This technique has been used extensively with various metal catalysts [77, 78].

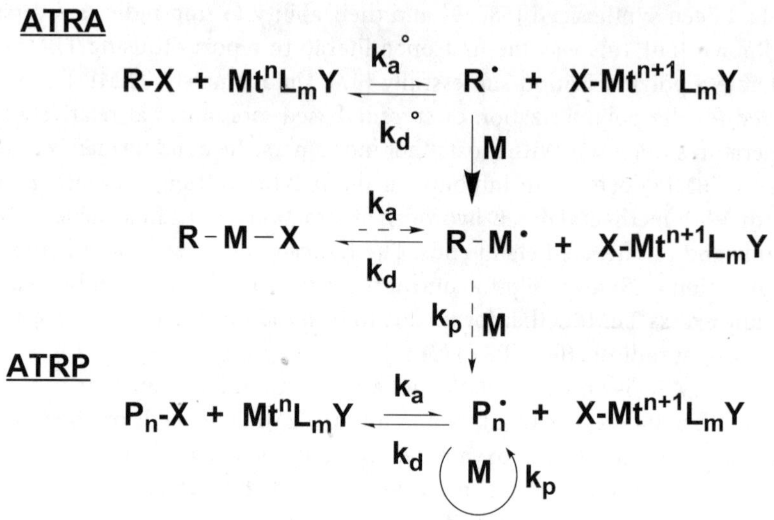

ATRA

$$R\text{-}X + Mt^nL_mY \xrightleftharpoons[k_d^{\circ}]{k_a^{\circ}} R^{\bullet} + X\text{-}Mt^{n+1}L_mY$$

$$\downarrow M$$

$$R\text{-}M\text{-}X \xrightleftharpoons[k_d]{k_a} R\text{-}M^{\bullet} + X\text{-}Mt^{n+1}L_mY$$

$$\downarrow k_p \, M$$

ATRP

$$P_n\text{-}X + Mt^nL_mY \xrightleftharpoons[k_d]{k_a} P_n^{\bullet} + X\text{-}Mt^{n+1}L_mY$$

$$(M) \, k_p$$

Scheme 4. ATRA and ATRP

To promote a polymerization, the newly formed carbon-halogen bond must be capable of being reactivated and the new radical must be able to add another alkene. This was accomplished for the radical polymerizations of St and methyl acrylate (MA), which were initiated by 1-phenylethyl bromide and catalyzed by a Cu(I)/2,2'-bipyridine (bpy) complex [42, 79–81]. The process was called "Atom Transfer Radical Polymerization" (ATRP) to reflect its origins in ATRA. A successful ATRP relies on fast initiation, where all the initiator is consumed quickly, and fast deactivation of the active species by the higher oxidation state metal. The resulting polymers are well defined and have predictable molecular weights and low polydispersities. Other reports used different initiator or catalyst systems, but obtained similar results [43, 82]. Numerous examples of using ATRP to prepare well-defined polymers can now be found [44–47, 49]. Scheme 4 illustrates the concepts of ATRA and ATRP. To simplify schemes 3, 4 and 5, termination was omitted.

1.3.3
Degenerative Chain Transfer Including RAFT

This technique for controlling radical polymerizations is based on one of the oldest technique, that of chain transfer, and has often been used in telomerization [83]. Similar to the concept of degenerative transfer with alkyl iodides [50, 51, 84], reversible addition fragmentation chain transfer with dithioesters (RAFT) [52–55, 85] is successful because the rate constant of chain transfer is faster than the rate constant of propagation. Analogous to both nitroxide-medi-

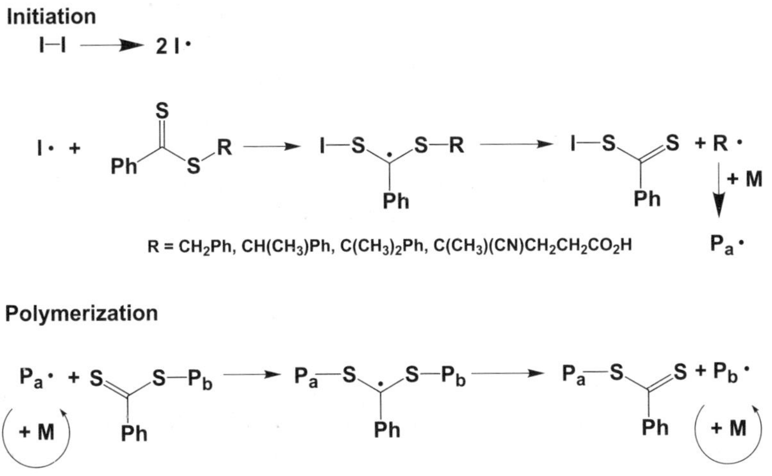

Initiation

I—I ⟶ 2 I·

I· + Ph-C(=S)-S-R ⟶ I-S-C·(Ph)-S-R ⟶ I-S-C(Ph)=S + R·
                                                              ↓ + M
R = CH$_2$Ph, CH(CH$_3$)Ph, C(CH$_3$)$_2$Ph, C(CH$_3$)(CN)CH$_2$CH$_2$CO$_2$H   P$_a$·

Polymerization

P$_a$· + S=C(Ph)-S-P$_b$ ⟶ P$_a$-S-C·(Ph)-S-P$_b$ ⟶ P$_a$-S-C(Ph)=S + P$_b$·
(+ M)   (+ M)

Scheme 5. The mechanism of RAFT polymerizations

ated and ATRP reactions, the polymer chains spend the majority of the reaction time in the dormant state and are only activated for a short period of time. Like iniferters, the RAFT agents are stabilized dithio compounds, which contain a small molecule capable of initiating a polymer chain. After homolytic cleavage to release the initiator, the RAFT agent can reversibly deactivate the polymer chains, resulting in a level of control over the polymerization not obtained with other chain transfer agents. Scheme 5 illustrates the concept of RAFT using a general radical initiator, I-I.

1.4
Summary

Each of the above methods for controlling the radical polymerization of vinyl monomers has its strengths and weaknesses. For example, the rates in ATRP can be easily adjusted through both the amount and activity of the transition metal complexes (both activator and deactivator). Faster rates in RAFT require larger amounts of initiators, i.e., more uncontrolled chains, while faster NMP requires less persistent radicals, which may result in more termination higher polydispersities. At the same time, transition metal complexes, although not attached to the polymer chains, require removal and can potentially be recycled.

Many ATRP initiators, including multifunctional systems, are either commercially available or very simple to make [86, 87]. Alkoxyamines and RAFT reagents are usually prepared from the corresponding alkyl halides. ATRP, however, requires the aforementioned catalysts, although they can be used in much less than equimolar amounts. The terminal halogens produced in ATRP can be easily converted to many useful functionalities, e.g., by nucleophilic substitution [88, 89]. Displacement of nitroxides and dithioesters is more difficult.

The range of polymerizable monomers is the largest for RAFT, but control requires adjustment of the dithioester structure and may be accompanied by retardation when targeting polymers with lower molecular weights [90]. NMP cannot yet be successfully applied to methacrylates. ATRP can be used to polymerize many monomers and, by using the halogen exchange, can be used for very efficient cross-propagation from acrylates to methacrylates, which is impossible to achieve by other methods [91].

Thus, each technique has some comparative advantages and limitations. Depending on the targeted material, (co)monomers used, range of molecular weights, composition, topology, and functionality, it may be advantageous to use one or another method. However, regardless of the drawbacks associated with each of these techniques, they have all been used extensively to prepare co-

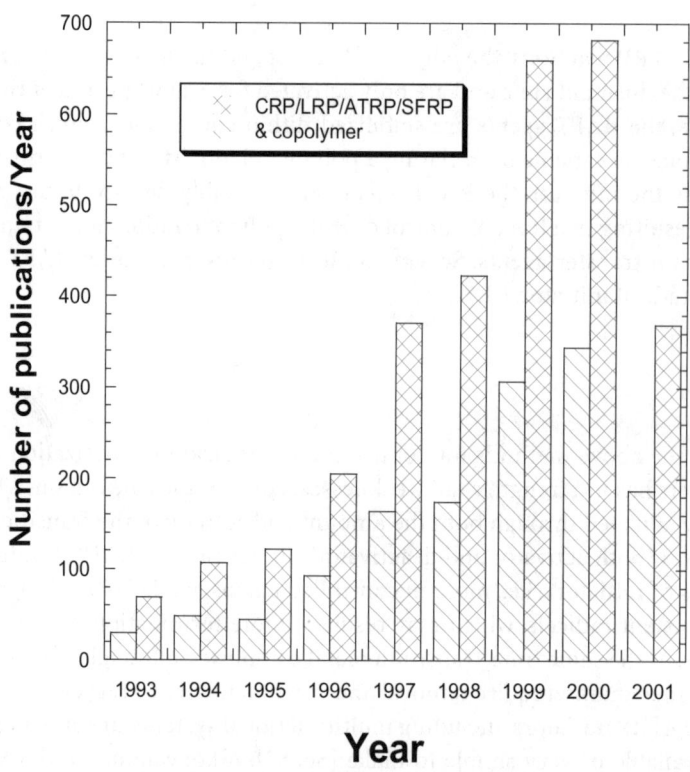

Fig. 3. The number of literature publications per year though August/2001 in the general field of controlled/"living" radical polymerization (CRP) and those related to specifically to CRP copolymers

polymers ranging from simple random copolymers to more complex densely grafted copolymers. The literature is filled with reports of novel chain architectures made directly, and indirectly, by CRP techniques. The discussion that follows is a detailed look at the types of copolymers prepared using NMP, RAFT, and ATRP-based reactions. The review covers the literature of the burgeoning field of modern controlled radical polymerization from 1993 through early 2001. The field of CRP develops extremely rapidly. Figure 3 presents the nearly exponential growth in the number of publications on controlled radical polymerization or living radical polymerization or atom transfer radical polymerization or stable free radical polymerization according to a search by SciFinder Scholar on August 15, 2001. Surprisingly, nearly half of these citations are in some way related to copolymers. Papers selected for this review were published primarily in peer reviewed journals.

2
Statistical Copolymers

The simplest type of copolymer is one where two or more comonomers are simultaneously copolymerized. This technique is commonly used to modify and/or improve the mechanical and physical properties of many polymers. The relative rates of incorporation of each monomer in a given set of comonomers will be dependent on their reactivity ratios. These have been determined for the radical polymerization of numerous monomer pairs [5]. Choosing comonomers that have reactivity ratios close to 1 produces statistical copolymers, where the radical chain ends react as often with their own monomer as they do with the others, assuming equal concentrations of both monomers. If both the reactivity ratios are significantly higher than 1, indicating that the radical would prefer to homopropagate rather than cross-propagate, the backbone copolymer exhibits a more blocky structure. If both are much lower than 1, indicating that both radicals would prefer to cross-propagate, the copolymer takes on an alternating structure. This is the most typical case for a radical polymerization. If the copolymerization obeys Bernoullian statistics, a random copolymer is formed. However, polymerization may follow other types of statistics (e.g., Markovian), resulting in a variety of statistical copolymers.

Gradient copolymers, also known as tapered copolymers, describe a chain topology that is generally unique to living polymerizations [6, 8]. They are intermediates between statistical and block copolymers, as shown in Fig. 4.

These copolymers result from copolymerizations where one active species would prefer to homopropagate and the other to cross-propagate, but neither tendency is extreme. The living requirement stems from fast initiation and the

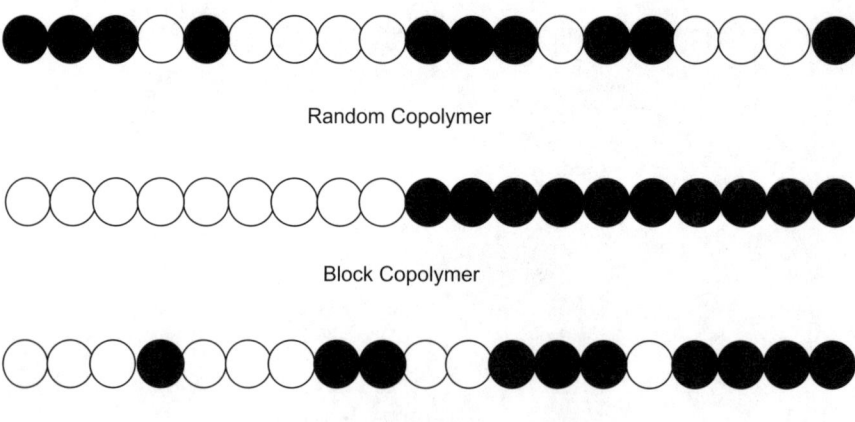

Fig. 4. Illustration of random, block, and gradient copolymers

need for an increased lifetime of the active species, which allows for many cross-propagations to occur during chain growth. For example, a gradient-type backbone forms in the radical copolymerization of methyl methacrylate (MMA) with *n*-butyl acrylate (nBA) [92, 93]. The reactivity ratios are 1.7 for MMA and 0.2 for nBA. With equimolar amounts of each monomer in the feed, the MMA will be preferentially incorporated during the early stages of the polymerization, but as the MMA is consumed, the composition of the nBA in the chain will increase. This results in a spontaneous gradient along the backbone where each end is enriched in each homopolymer but the middle is a tapered segment of both monomers. The properties of such copolymers are quite different from pure block copolymers [94]. The "breadth" of the gradient will be determined by the reactivity ratios of the comonomers. "Forced" gradient copolymers can also be prepared by continuous or periodic addition of one of the monomers [8].

2.1
SFRP/NMP

Hawker et al. [95] and Fukuda et al. [96] both reported on the copolymerization of St with various monomers in 1996. Hawker reported copolymerizations with nBA, MMA, and *p*-chloromethylstyrene (CMSt), while Fukuda focused on several acrylates, 9-vinylcarbazole, and acrylonitrile (AN), and succeeded in preparing block-random copolymers of St with AN (details below). Neither group found the polymerizations to be well-controlled when low concentrations of St were present in the comonomer feed. However, since then, NMP has been used extensively to prepare copolymers. Pozzo et al. copolymerized St with 4-vinyl pyridine (VP), initiated by benzoyl peroxide (BPO) and using TEMPO as the radical mediator [97]. After purification, the copolymer was reacted with spiro [fluorenecyclopropene] to prepare photochromic copolymers with controlled molecular weights (Scheme 6).

Expanding on their earlier work, Hawker et al. used the TEMPO system to synthesize both block and random copolymers of St and *p*-acetoxystyrene (AcOSt). They found that the rate of aqueous base dissolution was much greater for the random copolymers than for the corresponding block copolymers, which is of critical importance in microlithography [98]. Similar work by Yoshida used 4-methoxy-TEMPO (MOTEMPO) to copolymerize St with *p*-bromostyrene [99] or St with *p*-methoxystyrene followed by St with *p-tert*-butoxystyrene, and vice versa [100]. The presence of *p*-bromostyrene imparted flame resistance to the pSt [99] and hydrolysis of the *tert*-butoxy groups in the other copolymers led to vinyl phenols without affecting the methoxy groups, providing a route to vinyl phenol units in the polymer backbone.

The early work by Fukuda et al. on the St/AN systems had found that with an azeotropic mixture of St with AN, (63 mol% St in the feed), the polymerization

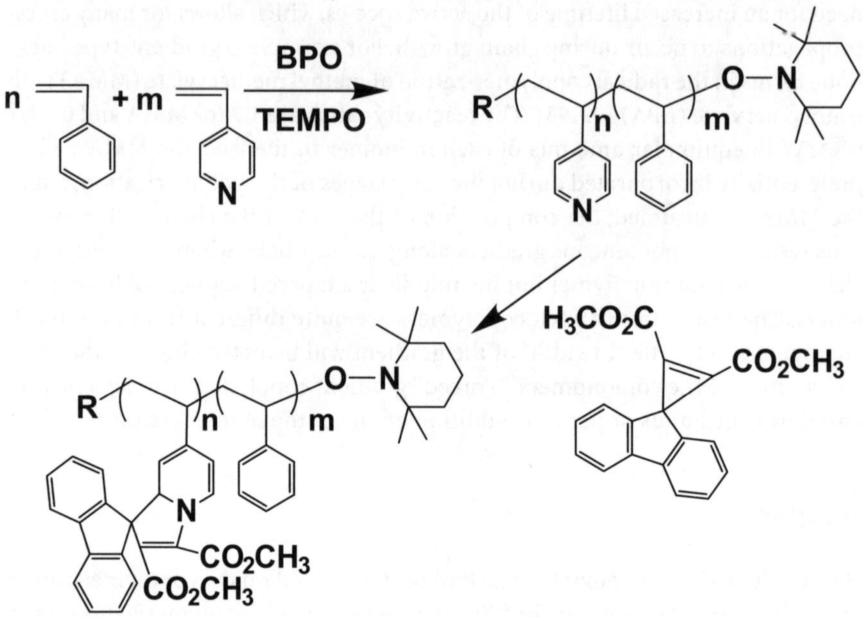

Scheme 6. Preparation of photochromic copolymers using a TEMPO-mediated polymerization [97]

was controlled, reaching 69% total monomer conversion in 10 h with M_n=16,000 and M_w/M_n=1.23 [96]. Similar control was obtained when a TEMPO-capped pSt macroinitiator was used for the copolymerization, resulting in a block-random copolymer with M_n=68,000 and M_w/M_n=1.30. Morphological characterization indicated the block copolymer formed a lamellar structure and in a selective solvent for pSt, micelles formed, providing evidence for the presence of chemically linked sequences rather than a blend of two different polymers. Baumert and Mülhaupt also synthesized copolymers of St with AN, however, they used 4,4'-azobis(4-cyanopentanecarboxylic acid) as the initiator, as opposed to the traditional peroxides, to impart a carboxy functionality onto the chain end [101]. The monomer feed ratio (73 wt% St/27 wt% AN) was reflected in the unchanging composition of the copolymer over the entire conversion regime, indicating that the polymerization was azeotropic. The final polymers had an M_n >17,000 with a PDI (M_w/M_n)=1.39. The sequence distribution of AN along the backbone was similar to that found in conventional free radical copolymerization, indicating the polymerization proceeded via a radical process [101].

Schmidt-Naake and Butz investigated the copolymerization of St with N-cyclohexylmaleimide (CMI) using the BPO/TEMPO system [102] and found that the rate of copolymerization was faster than the rate of either homopolymerization using the same system. They concluded that the electron-donating St and

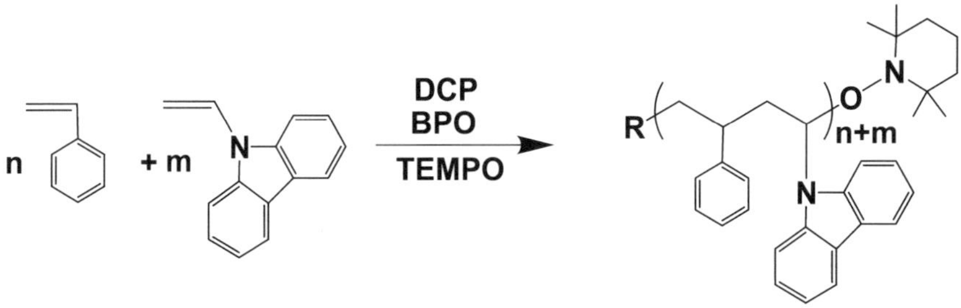

Scheme 7. Statistical copolymerization of St with NVC mediated by TEMPO [110]

the electron-withdrawing CMI formed a donor-acceptor type complex, which enhanced the rate of monomer addition [102]. In a later work by Lokaj et al., substituting n-butylmaleimide for CMI, resulted in similar findings, even when St was used to generate the initiating radicals through a thermal process instead of adding a small molecule initiator [103]. Alternating and block-alternating copolymers have recently been prepared from maleic anhydride and styrene using the more reactive nitroxides [104].

Schmidt-Naake et al. investigated the copolymerization of St and N-vinylcarbazole (NVC) initiated by BPO and mediated by TEMPO (Scheme 7) [105]. NVC cannot be homopolymerized with the TEMPO system and they found that, as the concentration of NVC in the comonomer feed increased, the rate of the polymerization decreased. However, in the presence of dicumyl peroxide (DCP), the rate of the polymerization was significantly enhanced compared to the system using BPO alone as the initiator. They concluded that adding the DCP to the polymerization reduced the concentration of free TEMPO in the system, analogous to when radicals are generated from the thermal self-initiation of St, resulting in an enhanced polymerization rate [105]. This conclusion is similar to those reached in earlier work by Mardare and Matyjaszewski [106] and Hawker et al. on the controlled autopolymerization of St in the presence of TEMPO [107] and by Greszta and Matyjaszewski [108]. Hawker et al. observed that in a copolymerization of St with MMA (which cannot be homopolymerized in a controlled way with the TEMPO system), the rate of polymerization decreased as the concentration of St in the comonomer feed decreased, and an incubation period began to occur, the length of which was directly related to the concentration of St in the feed. The molecular weight distributions also increased with a decrease in the concentration of St. Both systems were dependent on the presence of St to control the concentration of free TEMPO and thus regulate the rate of the polymerization.

In related work, Schmidt-Naake et al. used TEMPO-capped pSt to initiate the copolymerization of St and n-butyl methacrylate (nBMA) [109]. The authors found that the rate of the copolymerization was independent of the concentra-

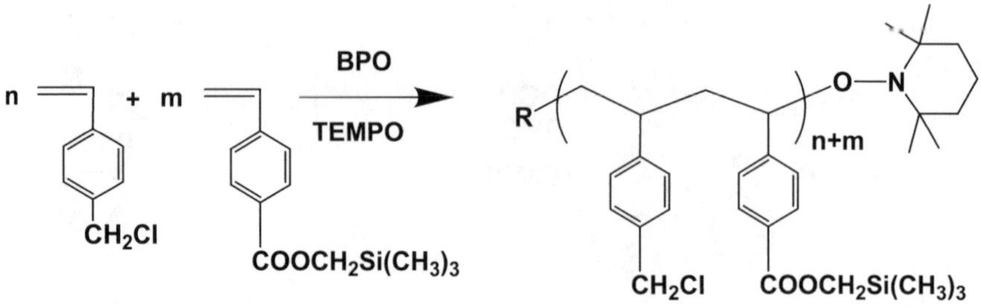

Scheme 8. Synthesis of copolymers for use as electron beam photoresists [112]

tion of the macroinitiator and was nearly identical to the rate of thermally self-initiated copolymerization of St/nBMA mixtures, in agreement with their earlier results [110], the work of Hawker et al. [107], as well as with the seminal work of Fukuda et al. [111]. However, although the authors claim that because the polydispersities were low and the molecular weights increased the polymerization was well controlled, the monomer conversion was limited to 20%, indicating the presence of irreversible chain breaking reactions, even with a low content of nBMA in the feed (St:nBMA=8:2) [109].

Nitroxide-mediated polymerizations can also be used to prepare copolymers that contain various additional functionalities. Ober et al. copolymerized CMSt with a silicon based styrenic monomer, trimethylsilylmethyl 4-vinylbenzoate (MVB-TMS), using the BPO/TEMPO system to prepare novel photoresists (Scheme 8) [112]. The monomer feed ratio of MVB-TMS:CMSt was held constant at 5:1 as was the total degree of polymerization at 200. The molecular weights were predictable, the polydispersities were low ($M_w/M_n<1.5$), and there was a marked improvement in the lithographic resolution using these copolymers [112]. Jones et al. attempted to prepare copolymers of St and epoxystyrene (EPSt), but the size exclusion chromatograms were bimodal, indicating that the polymer chains were not uniform [113]. Further investigation led them to conclude that at 124 °C, the temperature needed for the St/TEMPO system, the rate of thermal polymerization of epoxystyrene was high enough to result in polymers that were not controlled in the presence of TEMPO; better results were obtained by ATRP [113].

In general, TEMPO-mediated polymerizations have been successfully used to prepare copolymers of St-based monomers; however, attempts to incorporate other monomers have been difficult. The major reason behind this limitation is that radicals generated by the thermal self-initiation reaction of St are required to moderate the rate of polymerization by consuming the excess nitroxide produced by termination. When the ratio of St in the monomer feed is high, copolymerization with non-St based monomers is possible; however, as the level of St

Table 1. Summary of statistical copolymerizations performed in the presence of various TEMPO derivatives

Comonomers	TEMPO deriv.	Results	Investigator
St/nBMA	TEMPO	Controlled if monomer feed St rich	Hawker et al. [95], Schmidt-Naake et al. [109]
St/ClMS	TEMPO	Controlled if monomer feed St rich	Hawker et al. [95]
St/MMA	TEMPO	Controlled if monomer feed St rich	Hawker et al. [95]
St/AN	TEMPO	Can use as low as 63 mol% St in feed	Fukuda [96], Baumert and Mülhaupt [101]
St/NVC	TEMPO	Controlled if monomer feed St rich or additional radical source added	Fukuda et al. [96], Schmidt-Naake et al. [110]
St/VP	TEMPO	Controlled if monomer feed St rich	Pozzo et al. [97]
St/AcOSt	TEMPO	Controlled	Hawker et al. [98]
St/BrSt	MOTEMPO	Controlled	Yoshida [99]
St/MSt	MOTEMPO	Controlled	Yoshida and Takiguchi [100]
St/BuSt	MOTEMPO	Controlled	Yoshida and Takiguchi [100]
St/CMI	TEMPO	Controlled, fast rate of polymerization due to donor/acceptor complex formation	Schmidt-Naake and Butz [102]
St/BMI	TEMPO	Similar to St/CMI	Lokaj et al. [103]
CMSt/MVB-TMS	TEMPO	5:1 ratio of MVB-TMS:CMSt, controlled	Ober et al. [112]
St/MAh	TIPNO	Alternating and block/alternating copolymers	Hawker et al. [104]
St/EPSt	TEMPO	Loss of end groups, rate of polymerization of EPSt too fast at polymerization temperature	Jones et al. [113]

in the feed decreases, the rate of polymerization decreases and the comonomer consumption is incomplete. Table 1 contains a summary of the types of statistical copolymers prepared using various TEMPO-based systems, as well as general information about the polymerization and the corresponding references. Future development of new nitroxides should increase the usefulness of this method for copolymerizations.

2.2 ATRP

In 1995, Matyjaszewski et al. used ATRP to prepare gradient copolymers of St with MA using a CuCl/bpy catalyst, and with MMA using a CuBr/bpy catalyst

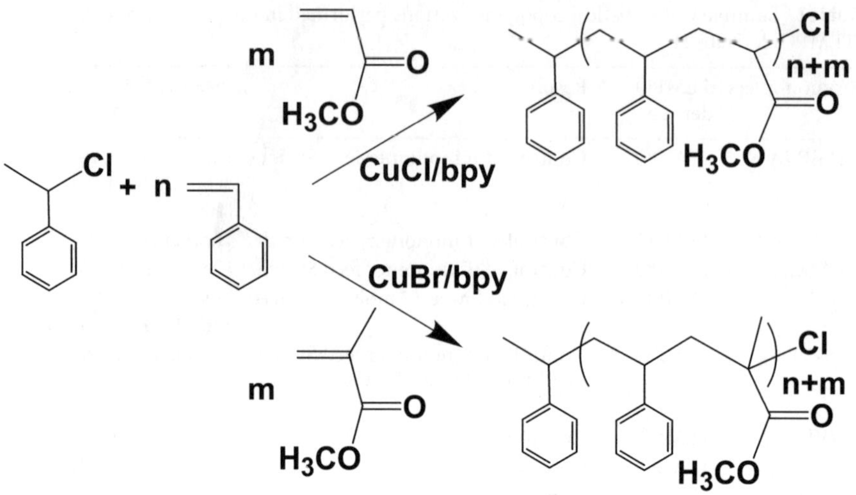

Scheme 9. Copolymers formed using ATRP methods [114]

system (Scheme 9) [114]. The copolymerization of a 50:50 molar ratio of St and MA showed a linear increase of molecular weight with increasing conversion and a narrow final molecular weight distribution (M_w/M_n=1.25). In the St/MMA system, the content of St in the copolymer decreased with an increasing rate of MMA addition, suggesting the formation of a forced gradient along the backbone [114].

Subsequently, gradient copolymers of MMA with MA were prepared via a one-pot copolymerization as well as a gradient copolymer of St with MA using a monomer addition technique [6]. The shape of the gradient in the St/MA system was altered by changing the rate of MA monomer addition from 0.10 ml/min, which produced a weak gradient, to 0.05 ml/min, which resulted in an almost blocky type structure in the backbone [6]. Thermal and mechanical studies on the block, statistical, and forced gradient copolymers of styrene and methyl acrylate with molecular weights M_n~10,000 to 20,000, polydispersiies M_w/M_n<1.2, and compositions ~50% are shown in Fig. 5. The DSC and dynamic mechanical results indicate that the block copolymer has significantly different properties from the gradient copolymers. The DSC traces show that the forced gradient copolymer behavior depends on the thermal history. The lower modulus (G') of the forced gradient compared to the statistical gradient copolymer demonstrates that materials with different properties were produced.

Gradient copolymers of St and AN were also prepared [115, 116]. Since St and AN have reactivity ratios that are both significantly less than 1, the copolymerization has an alternating tendency along the backbone with an enrichment in AN at high conversions. Alternatively, a gradient along the backbone can be

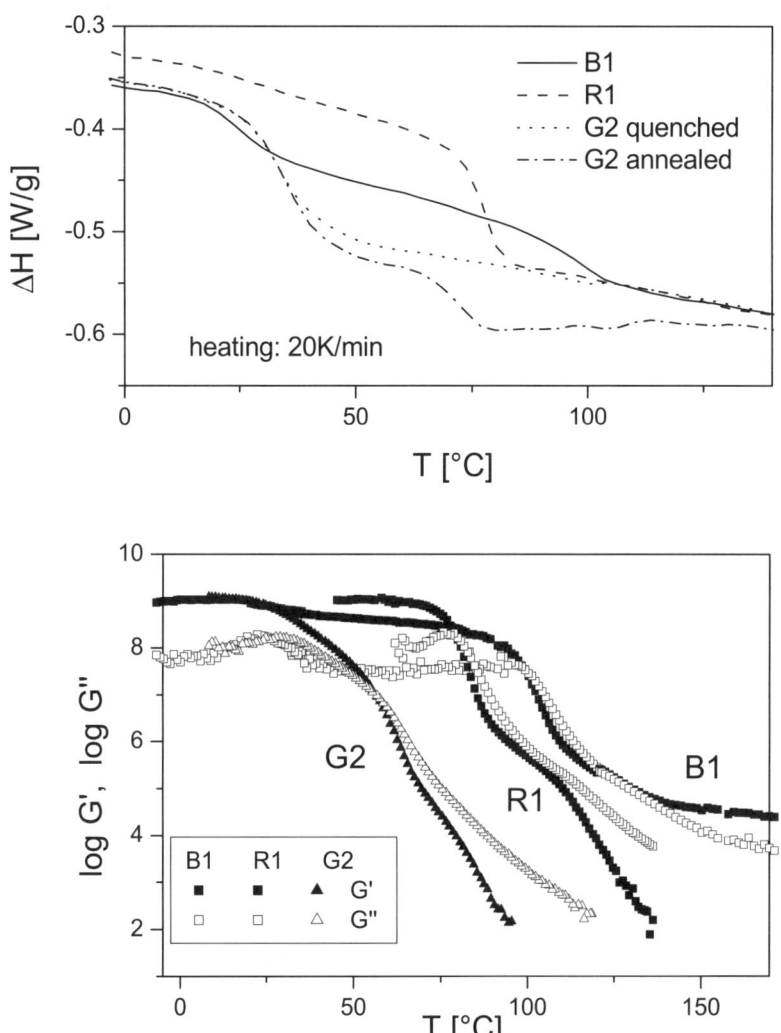

Fig. 5. DSC (above) and dynamic mechanical results (below) for copolymers of St with MA; B=blend, R=random, G=gradient, $M_n \sim 10,000$ to $20,000$, $M_w/M_n < 1.2$, compositions ~50% [6]

forced to occur by adding the second monomer gradually throughout the course of the polymerization. The shape of the gradient was altered by changing the rate of addition of the second monomer [115]. The gradient copolymers had thermal and mechanical properties that were significantly different from the corresponding statistical and block copolymers [116]. This indicates that these types of copolymers, with a controlled composition along the backbone but with a homogeneous composition among the chains, show promise as a new class of materials with tunable properties. Figure 6 shows the results of small angle X-ray scattering of two low polydispersity S/AN gradient copolymers with the same content of AN (59 mol%), but different MW (M_n=11,000, M_w/M_n=1.15; M_n=25,000, M_w/M_n=1.08). The higher MW sample displayed a periodicity of 22.4 nm, but the lower MW sample had only a periodicity of 13.4 nm. The higher MW sample stayed in the phase separated regime at T>200°C, but the order in the lower MW sample decreased with increasing temperature and a single phase was formed at T>150°C. Thus, it is possible to manipulate phase transitions by changing the MW and, perhaps, the compositions and the shape of the gradient.

Figure 7 shows temperature dependencies of the storage (G') and loss moduli (G") for both styrene/acrylonitrile gradient copolymers described above. For comparison, results for a random copolymer are also shown (M_n=39,000, M_w/M_n=1.08, ϕ_{AN}=0.45).

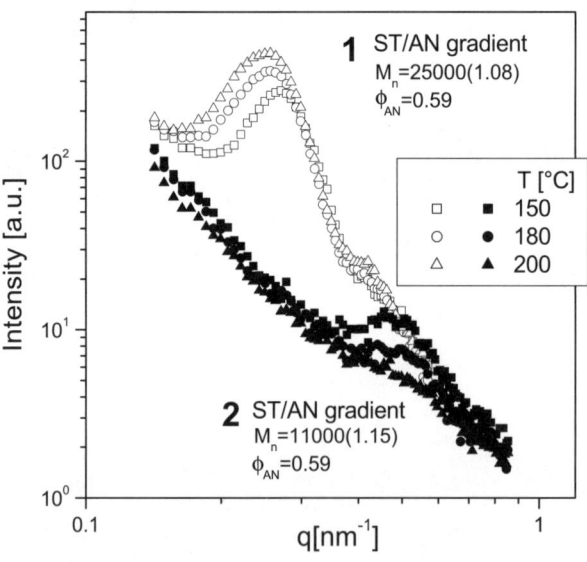

Fig. 6. Small angle X-ray scattering of two S/AN gradient copolymers containing. 59 mol% AN; gradient 1: M_n=25,000 and M_w/M_n=1.08; gradient 2: M_n=11,000 and M_w/M_n=1.15 [115]

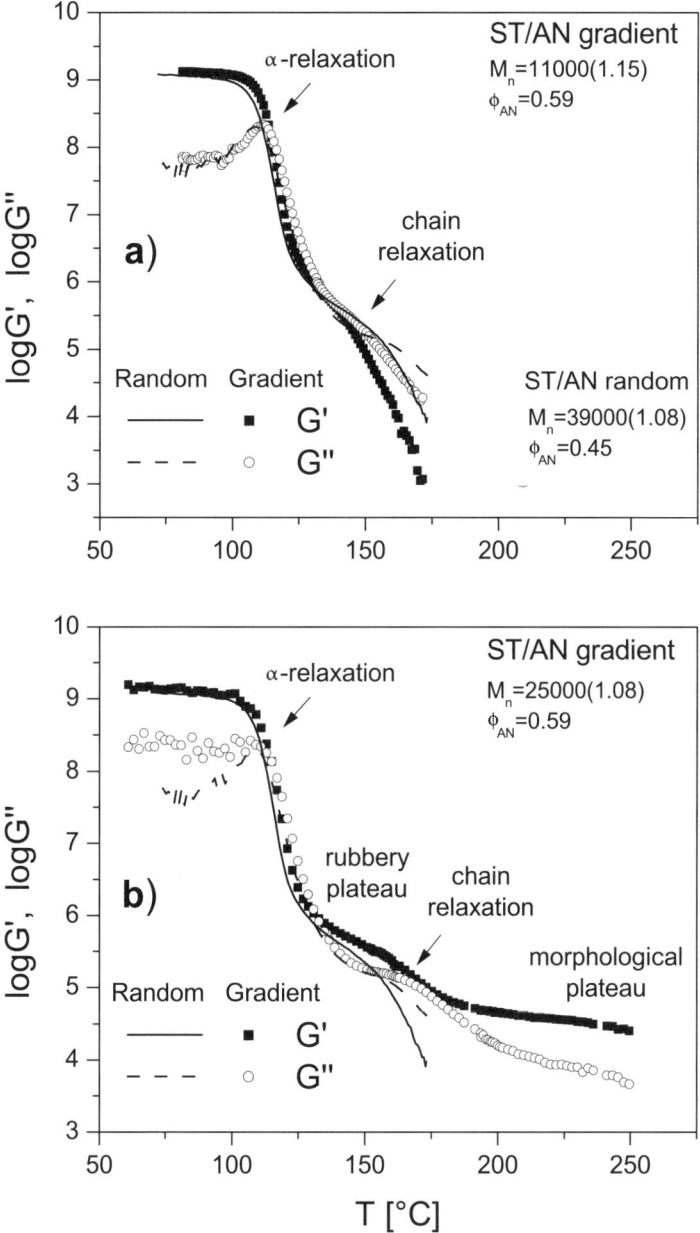

Fig. 7a,b. Temperature dependencies of the storage (G′) and loss moduli (G″) for St/AN gradient copolymers containing 59 mol% acrylonitrile: a) M_n=11,000, M_w/M_n=1.15; b) M_n= 25,000, M_w/M_n=1.08; random copolymer: M_n =39,000, M_w/M_n=1.08, f_{AN}=0.45 [115]

Several other reports on copolymerizations using the copper-based ATRP systems have been reported. Haddleton et al. investigated the MMA/nBMA system and showed that the reactivity ratios of the two monomers supported a radical mechanism [117]. Arehart and Matyjaszewski investigated the copolymerization of St with nBA, initiated by 1-phenylethyl bromide and catalyzed by a CuBr/4,4'-di-(5-nonyl)-2,2'-bipyridine (dNbpy) system [118], and showed that the rate of the reaction progressively decreased with increasing monomer conversion. This could be attributed to the decomposition of the pSt chain ends in the presence of the relatively polar nBA monomer as well as to differences in the equilibrium constants for ATRP of St and nBA. The reactivity ratios and the tacticity of the final copolymers were similar to those found in a free radical processes, again indicating that the process was radical in nature [118]. Similar results were obtained by Chambard and Klumperman [119]. Vairon et al. used a slightly different system, chlorodiphenylmethane as the initiator in conjunction with a CuCl/bpy catalyst. This system is homogeneous in the presence of small amounts of N,N-dimethylformamide, and was used to copolymerize St with nBA. They found that the copolymer formed from a 1:1 molar ratio of St with nBA had a single glass transition temperature (T_g) at 26 °C, different from the two T_gs observed for a block copolymer [120].

The copolymerization of MMA with nBA was also studied using different catalytic systems. The calculated reactivity ratios were close to those for a conventional radical polymerization and similar for different Cu-based catalytic systems (bpy, PMDETA, and Me_6TREN) [92]. The distribution of triads and the polymer stereochemistry was as in any other free radical system [93].

The same monomer pair, as well as MMA/nBMA, were simultaneously copolymerized under heterogeneous conditions in water. Statistical copolymers with low polydispersities ($M_w/M_n<1.25$) were prepared in high yield [121, 122].

The unsuccessful attempts to copolymerize St with EPSt using a nitroxide-mediated reaction led Jones et al. to investigate ATRP as an alternative [113]. They used the CuBr/bpy catalyst system in conjunction with methyl 4-(bromomethyl)benzoate as the initiator. The copolymers had monomodal GPC curves and narrow molecular weight distributions ($M_w/M_n<1.25$) when the monomer feed contained <10% of the epoxystyrene and the polymerization temperature was kept at 100 °C. This is in contrast to the results obtained from the TEMPO-mediated polymerization where all the copolymers had bimodal GPC traces and broad molecular weight distributions [113].

There have been several other well-defined random copolymers based on styrene derivatives prepared with trimethylsilylstyrene [123], p-acetoxystyrene [124], p-methoxymethylstyrene, and p-acetoxymethylstyrene [125]. The resulting copolymers served as precursors to functional materials with Si, phenol, or hydroxybenzyl moieties, or for subsequent crosslinking or grafting processes.

Sawamoto et al. used the $RuCl_2(PPh_3)_3/Al(OiPr)_3$ catalyst to prepare St/MMA copolymers [126]. They found that the polymerization proceeded well using 1-phenylethyl bromide as the initiator and that the composition of the copolymer matched the comonomer feed composition, or behaved azeotropically [126]. The polymers were well-defined, with predictable molecular weights and relatively low polydispersities ($M_w/M_n<1.5$). The reactivity ratios were similar to those determined from conventional free radical processes. Later work used a $NiBr_2(n-Bu_3P)_2$ catalyst system for the ATRP of a 50/50 mixture of MMA/MA and MMA/nBA [127]. The results indicated that the copolymerization was controlled with copolymer $M_n=11,800$ ($M_w/M_n=1.47$) and 12,500 ($M_w/M_n=1.47$), respectively.

Using Wilkinson's catalyst, $RhCl(PPh_3)_3$, Jerome et al. reported a successful copolymerization of MMA with 2-hydroxyethyl methacrylate (HEMA) [128]. The polymerization was carried out without the addition of the Lewis acid $Al(OiPR)_3$ that was necessary for the ruthenium system and showed that the system was tolerant to the hydroxy functional groups, as expected for a radical process [128]. Demonceau et al. later modified the ruthenium system and used $RuCl_2(p\text{-cymene})(PR_3)$ (R=alkyl group that can be adjusted for each monomer) for the copolymerization of MMA with HEMA, as well as with methacrylic acid (MAA), further showing that adjusting the ligands can eliminate the need for added Lewis acid [129].

Alternating copolymers belong to a general class of periodic copolymers. They can be formed from a comonomer pair which strongly prefer crosspropagation over homopropagation. This is the case of electron rich and electron poor pair, such as N-phenyl and N-cyclohexyl maleimides/styrenes as well as isobutene/AN and MA systems [86, 130, 131]. Attempts to incorporate maleic anhydride (MAh) failed, apparently due to catalyst poisoning. Recently, Li et al. reported on preparation of alternating copolymers of St with different substituted maleimides via ATRP [132]. A CuBr/bpy catalyst system was used to polymerize a 50/50 molar ratio of St with N-(2-acetoxyethyl) maleimide (AEMI) or with N-phenylmaleimide (PMI) at 80 °C (Scheme 10). Both polymerizations followed first-order kinetics in monomer consumption, indicating a constant concentration of active species throughout the polymerization; however, the rate of po-

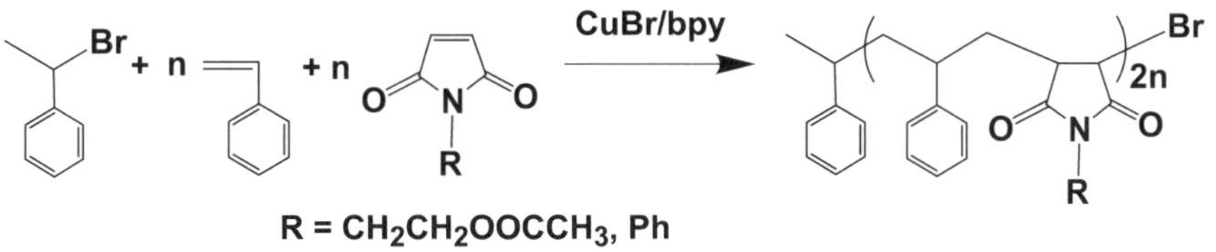

$R = CH_2CH_2OOCCH_3$, Ph

Scheme 10. Alternating copolymers of St with maleimide monomers prepared by ATRP [132]

Table 2. Summary of statistical copolymerizations performed using ATRP systems

Comonomers	Catalyst	Comments	Investigator
St/MA	CuCl/bpy	50:50 mole ratio, $M_w/M_n<1.3$	Matyjaszewski et al. [114]
St/MMA	CuBr/bpy	St content decreases with increasing monomer conversion, tapered	Matyjaszewski et al. [6, 114]
MMA/MA	CuBr/bpy	MA content increases with increasing monomer conversion	Greszta and Matyjaszewski [6]
St/AN	CuBr/bpy	Forced gradient, different mech. properties than statistical copolymer	Matyjaszewski et al. [115, 116]
MMA/nBMA	CuBr/PCPI[a]	r_1 (MMA)=0.98, r_2 (nBMA)=1.26	Haddleton et al. [117]
St/MMA	$RuCl_2(PPh)_3$/ $Al(OiPr)_3$	Azeotropic copolymerization	Sawamoto et al. [126]
MMA/BA	CuBr/bpy, TREN and PMDETA	Similar reactivity ratios to conventional RP, similar sequences and tacticity	Ziegler and Matyjaszewski [92], Madruga et al. [93]
MMA/BA; MMA/MA	$NiBr_2(P\,n\text{-}Bu_3)_2$/ $Al(Oi\text{-}Pr)_3$	50:50 MMA:BA or MMA:MA, $M_n<13,000$, $M_w/M_n<1.5$	Sawamoto et al. [127]
MMA/HEMA	$RhCl(PPh_3)_3$	Well-controlled	Jerome et al. [128]
MMA/HEMA	$RuCl_2/(p\text{-}cymene)(PR_3)$	90:10 MMA:HEMA, controlled	Demonceau et al. [129]
MMA/MAA	$RuCl_2/(p\text{-}cymene)(PR_3)$	95:5 MMA:MAA, controlled	Demonceau et al. [129]
St/nBA	CuBr/dNbpy	Reactivity ratios as expected for radicals	Matyjaszewski [118]; Klumperman [119]
St/nBA	CuBr/bpy	1:1 nBA:St, single T_g	Vairon et al. [120]
St/EPSt	CuBr/bpy	<10% EPSt controlled	Jones et al. [113]
MA/VOAc	CuBr/bpy	<30% of VOAc incorporated, $M_n=11,140$, PDI=1.16	Matyjaszewski et al. [86]
St/pAcOSt	CuBr/bpy	Azeotropic copolymerization	Kops et al. [124]
St/pMeOMeSt;St/pMeOAcSt	CuBr/bpy	For grafting after deprotection	Doerffler and Patten [125]
St/Sty-TMS	CuBr/bpy	Azeotropic copolymerization	McQuillan et al. [123]
St/PMI;AN/IB; MA/IB	CuBr/bpy	Alternating copolymers, $M_n=4730$, PDI=1.19	Matyjaszewski et al. [86]
St/AEMI or St/PMI	CuBr/bpy	Alternating structure along the backbone	Li et al. [132]

[a] 2-Pyridine-carbaldehyde n-propylimine

lymerization of the AEMI was faster than the PMI system, and this was attributed to differences in the solubilities of the catalyst system [132]. There was a linear increase of the molecular weight with conversion in both systems, indicating negligible transfer. Analysis of the composition of the copolymer confirmed the alternating structure of the backbone over a large range of monomer feed conditions.

ATRP is a useful tool for preparing statistical copolymers with various monomer combinations. Unlike the TEMPO systems detailed above, the ATRP systems can be used to copolymerize styrene, acrylate, or methacrylate based combinations, potentially leading to materials with better and/or different physical and mechanical properties than the corresponding homopolymers or block copolymers. This may also include monomers which cannot yet be homopolymerized by ATRP such as isobutene or vinyl acetate [86, 130]. Table 2 summarizes statistical copolymers prepared using ATRP systems.

2.3
Degenerative Transfer Processes

RAFT has also been used to prepare copolymers. The copolymerization of MMA with nBA in the presence of cumyl dithiobenzoate as the transfer agent resulted in a polymer with a gradient of composition along the backbone, well-defined molecular weights, and low polydispersities [53]. Several copolymers were made by degenerative transfer with alkyl iodides [133].

Interesting alternating copolymers and alternating/block copolymers were also prepared from styrene and maleic anhydride [134]. It has been earlier reported for conventional radical systems that tendency for alternation can be enhanced in the presence of Lewis acids, e.g., using $EtAlCl_2$ in St/MMA system [135–137]. Using the same Lewis acid in addition to a RAFT reagent, Matyjaszewski et al. obtained a strongly alternating copolymer between St and MMA with a low polydispersity [138].

Simultaneous copolymerization of MMA with MMA terminated poly(dimethylsiloxane) macromonomers results in the formation of a gradient of composition of PDMS units along the backbone due to a lower macromonomer reactivity which additionally decreases at lower temperatures [139].

2.4
Comparison of Various CRP Methods Applied to Statistical Copolymers

Statistical copolymers are an important class of materials, as their properties can differ significantly from the corresponding homopolymers or block copolymers. The major problem, prior to the advent of CRP methods, was that with conventional radical processes, slow continuous initiation produced a compositional drift among the chains and mixtures of the different copolymers and homopolymers were obtained. With CRP methods, the nearly simultaneous growth of all the polymer chains produces a homogeneous composition among the polymer chains, providing a route to a structure/property correlation. All CRP methods can be used to prepare these types of copolymers; however, each has distinct advantages and disadvantages. The reversible activation process in ATRP may alter some regio, stereo and chemoselectivity. In copolymerization, differences in activation/deactivation will effect relative concentration of active centers and overall rates. They should have minimal effects on relative rates of monomer consumption, especially, if comonomers tend to alternate. However, for reactivity ratios >1, faster activation of the dormant species derived from one monomer should lead to its faster incorporation into polymer chain, especially at low conversion when the crosspropagation equilibrium is not yet established. A similar effect will be observed if one monomer reacts with the initiator faster than the other one. This may provide apparent reactivity ratios, different from those in the process without reversible activation. Thus, low conversion data in copolymerization in ATRP and other CRP processes are less reliable for the reactivity ratios determination. Additional complications may be due to the potential complexation of comonomers with transition metal complexes.

The TEMPO system can be used for systems containing some St, which is required to moderate the rate of polymerization through spontaneous formation of radicals that consume any excess TEMPO. The disadvantage is that St must be present in order to use TEMPO or the addition of a radical "source" is necessary (i.e., DCP [108]). New nitroxides have overcome this limitation, and can be used to incorporate dienes and acrylates into (co)polymers, but methacrylates remain a challenge (see below). Little information about copolymers prepared by RAFT is presently available, but it shows promise as a technique for incorporating difficult monomers like vinyl acetate into polymers prepared by CRP methods. ATRP is versatile and robust since the catalyst system can be tuned to accommodate the desired monomer combinations through choice of both the transition metal and the ligand. Some systems may not be well-suited to specific comonomer combinations, however, due to strong differences in the equilibrium constants, which may produce too much deactivator, or too low a concentration of the active species, and therefore lead to incomplete monomer consumption. This can also occur in the nitroxide systems. Although the ATRP catalyst systems are generally robust and can tolerate several different types of impurities, some monomers (especially acids) may poison the catalyst and no polymerization occurs. This is less of a concern with either the RAFT or nitroxide systems. In general, the CRP methods can easily be used to prepare statistical and gradient copolymers. TEMPO can be used for systems containing St, while newer nitroxides can be used for acrylates and dienes. ATRP can be used for acrylates, methacrylates, styrenes, and acrylonitrile with the choice of catalyst depending on the comonomer system. VOAc, IB, and other comonomers not yet homopolymerizable by ATRP can also be incorporated via other methods.

3
Linear Block Copolymers

There have been numerous literature reports on the preparation of block copolymers using CRP methods. These copolymers range from those synthesized wholly by CRP to those that involve either transformation from other living polymerization techniques (anionic, cationic, ring-opening, etc.) to CRP, or functionalization of a macromolecule that can then be used as a macroinitiator for CRP. Each of these methods will be addressed separately. Nitroxides were predominantly used for styrene containing copolymers, whereas ATRP was successful for the acrylates and methacrylates as well [42].

3.1
Linear Block Copolymers Prepared Exclusively by CRP Methods

3.1.1
SFRP/NMP

The following sections detail the literature reports pertaining to the synthesis of block copolymers using nitroxide-mediated polymerization techniques. The sections are organized according to monomer type and generally follow the historical development of the particular subsection. Most literature on nitroxide mediated preparation of block copolymers is found for the styrene-based monomers, and is summarized first. This is followed by acrylates and dienes, as they were the next monomers to be studied. These sections are followed by more recent work with vinyl pyridine, acrylamides, and maleic anhydride. The final section deals with methacrylates. This is presented last to stress the importance of developing new nitroxides that can successfully be used for the homopolymerization of methacrylate-based monomers.

3.1.1.1
Styrene-Based Monomers

The majority of the block copolymers initially prepared using nitroxide-mediated polymerizations were based on styrene derivatives, simply because they were compatible with a TEMPO based system. Bertin and Boutevin conducted the CRP of CMSt in the presence of BPO and TEMPO to form a polymer which was chain extended with St to yield a block copolymer with a number average molecular weight, M_n=72,000 and a rather high polydispersity, M_w/M_n=1.8 (Scheme 11) [140]. The authors concluded that the blocking efficiency, which corresponds to the extent of functionality in the macroinitiator, was only 83%, resulting in a copolymer contaminated with a significant proportion of pCMSt homopolymer [140]. Yoshida and Fujii prepared various chlorostyrenes (CSt)

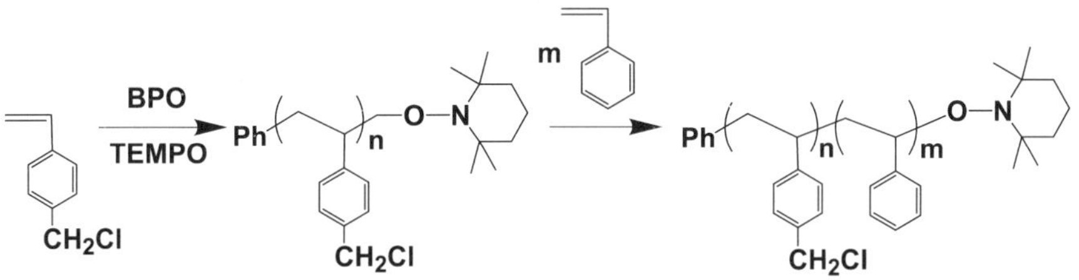

Scheme 11. Preparation of block copolymers of CMSt and St using TEMPO

homopolymers with the intention of using them as macroinitiators for a St polymerization; however, they utilized MOTEMPO instead of TEMPO [141]. They determined that the degree of polymerization was predictable based on the initial ratio of monomer to MOTEMPO and that the blocking efficiency decreased based on the position of the chloro substituent in the following order: 2>3>4. With p(4-CSt), bimodal GPC curves were observed [141], in accordance with the low blocking efficiency found by Bertin and Boutevin [140]. Yoshida attributed this to decomposition of the MOTEMPO chain end in poly(4-CSt). Later work by Boutevin et al. suggested that transfer reactions occurred between the TEMPO and the chloromethyl groups, thereby broadening the molecular distributions. However, when benzyl chloride was used as a model, there were no significant byproducts detected in the TEMPO-mediated polymerization of St to support this premise [142]. Nevertheless, they used pCMSt as the macroinitiator for the polymerization of St, and obtained no improvement over their earlier results [140].

In related work, Yoshida also prepared block copolymers of p-bromostyrene (BrSt) and St using MOTEMPO as the radical mediator [99]. Several low molecular weight pBrSt macroinitiators were prepared for the preparation of block copolymers with St, resulting in the formation of block copolymers with significantly higher molecular weights than the macroinitiator. However, the GPC traces were bimodal for two out of the three block copolymers, indicating there was a significant proportion of dead macroinitiator, about 10% based on the author's calculations [99]. Block copolymers prepared with the reverse order of blocks showed a significant increase in the polydispersity, from 1.14 to 1.34 upon addition of the BrSt, indicating the second block was not as well controlled as the first [99].

Catala et al. [143] prepared block copolymers of St with a substituted St, p-tert-butylstyrene (tBuSt); however, they used the 1-phenylethyl adduct of a slightly different nitroxide, di-tert-butyl nitroxide, to mediate the polymerization (A-T, Fig. 8) [63, 144]. This nitroxide allowed the polymerization to occur under milder conditions than normal for TEMPO-mediated reactions (90 °C vs

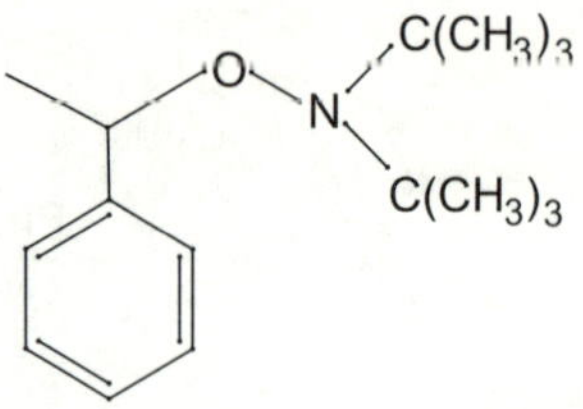

Fig. 8. Structure of A-T [63, 144]

120 °C). However, when this system was used for the homopolymerization of tBuSt, the monomer conversion had to be limited to <30% to avoid the formation of an insoluble network [143]. When pSt chain ends, capped with the di-*tert*-butyl nitroxide, were used to initiate the polymerization of tBuSt, chain extension resulted in well-defined copolymers with predictable molecular weights and narrow molecular weight distributions (M_w/M_n=1.2–1.3) [143]. Fukuda et al. also used nitroxide-mediated polymerizations to incorporate substituted styrenes into block copolymers, in particular *p-tert*-butoxystyrene (tBOSt) [145]. When 2-benzoyloxy-1-phenylethyl TEMPO was used as the unimolecular initiator, the authors found the polymerization was controlled and the resulting pt-BOSt could be used successfully as a macroinitiator for a St polymerization. Subsequent acid hydrolysis of the *tert*-butoxy groups led to block copolymers of poly(*p*-vinyl phenol)-*b*-polystyrene, which microphase separated to produce lamellar and cylindrical morphologies [145]. This work preceded Yoshida and Takiguchi's investigation of the random copolymers incorporating tBOSt discussed in the previous copolymer section [100].

Laus et al. also investigated a chain extension of pSt-TEMPO with a substituted styrene, phthalimide methylstyrene (PIMS) [146], with the objective of incorporating the PIMS, then deprotecting it to produce the amino functional polymers, according to Scheme 12.

Both AB and ABA triblock copolymers were targeted. When higher molecular weight pSt macroinitiators were used (DP>100), AB diblock was contaminated by the formation of homopolymer of PIMS, presumably by a thermally initiated polymerization mechanism. However, when a lower molecular weight macroinitiator was used (DP=38), the rate of formation of the block copolymer was faster than the rate of the thermally initiated polymerization, effectively eliminating formation of the homopolymer of PIMS, and leading to successful synthesis of pure block copolymer [146]. These AB block copolymers were then chain extended further with St; however, the GPC traces were bimodal, indicating a significant portion of dead chains. Nevertheless, the deprotection reaction was carried out on both the AB and ABA block copolymers producing blocks with amino-functionality, which was confirmed through IR analysis [146].

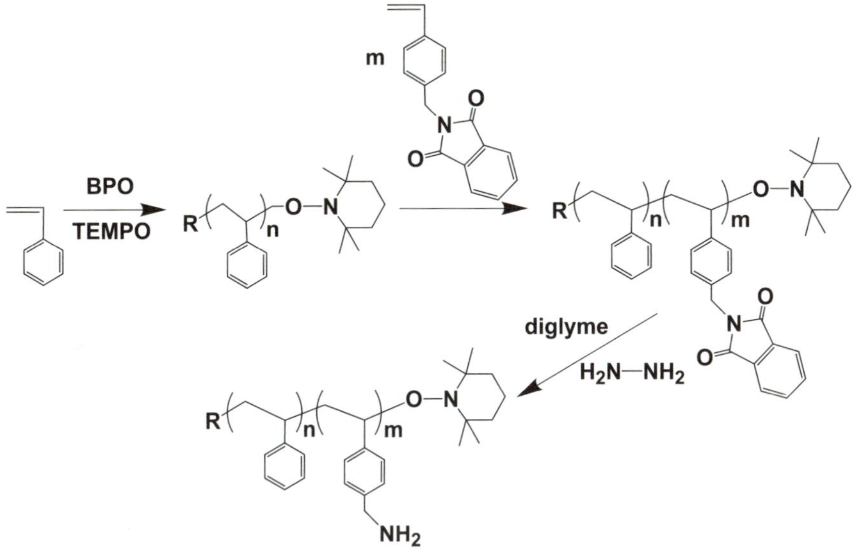

Scheme 12. Preparation of amino-functional block copolymers using TEMPO-mediated polymerization [146]

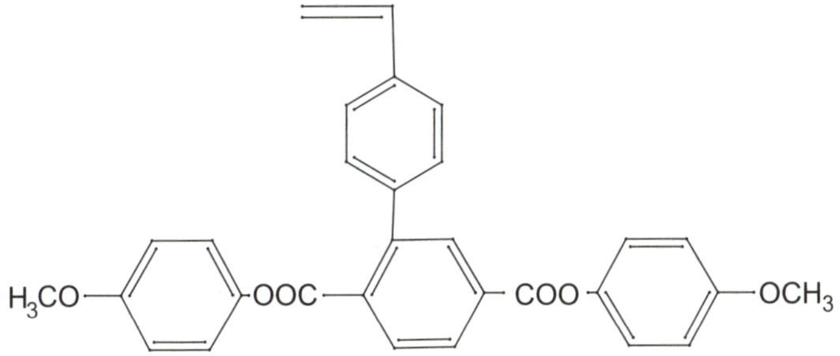

Fig. 9. 2,5-bis [(4-Methoxyphenyl)oxycarbonyl] styrene (MPCS) [148]

Wan et al. used TEMPO-mediated polymerizations to prepare liquid crystalline (LC) polymers [147, 148]. pSt-TEMPO was chain extended with a mesogen-jacketed LC monomer, 2,5-bis [(4-methoxyphenyl)oxycarbonyl] styrene (MPCS, Fig. 9) to form a rod-coil diblock copolymer. The resulting copolymer had an $M_n=19{,}500$ with an $M_w/M_n=1.48$. There was tailing to lower molecular weights, indicating the presence of some unreacted macroinitiator, but after extraction with cyclohexane, the remaining macroinitiator was removed, leaving pure block copolymer. ^{1}H and ^{13}C HMR analysis indicated the presence of both blocks, as did DSC analysis, which showed two T_gs, one at 117.2 °C (pMPCS) and

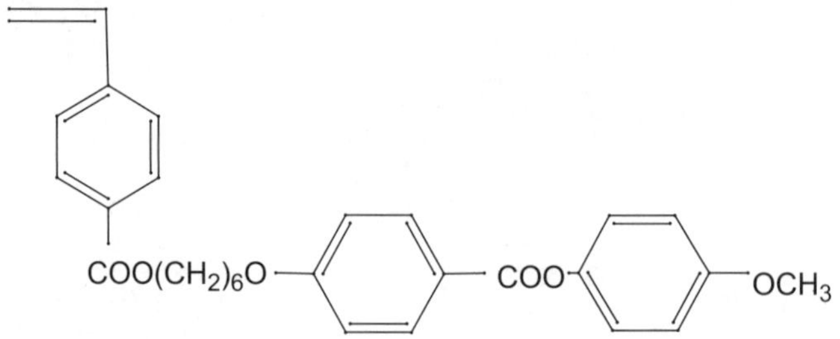

Fig. 10. [(4′-Methoxyphenyl)4-oxybenzoate] -6-hexyl (4-vinylbenzoate) (MPVB) [149]

the other at 93.2 °C (pSt) [148]. In contrast to the authors experience with a random copolymer of St with MPCS, the molar content of the MPCS in the block copolymer needed to display LC behavior was significantly lower (27–37% vs 79%), suggesting that the close proximity of the bulky side groups in the blocks may force the polymer chain into an extended conformation, thereby producing the desired rod-coil structure [148].

Other LC-based copolymers incorporating styrene-based monomers were prepared by Ober et al. [149] who chain extended pAcOSt-TEMPO (M_n=7000, M_w/M_n=1.18) with [(4′-methoxyphenyl)4-oxybenzoate]-6-hexyl (4-vinylbenzoate) (MPVB, Fig. 10). The reactions were controlled, with molecular weights ranging from M_n=12,600–23,000 and M_w/M_n=1.19–1.44. The content of pMPVB in the copolymer determined by ^{1}H NMR increased as the molar ratio of the MPVB to pAcOSt-TEMPO increased [149]. For two out of the three copolymers prepared a smectic-isotropic transition was observed; however, it was at a value lower than expected based on the composition of the copolymer, even after annealing. X-Ray diffraction patterning indicated that the copolymer was oriented in a lamellar morphology and that the smectic layers were perpendicular to the block copolymer lamellae [149].

The copolymers of St with CMI prepared by Schmidt-Naake et al., detailed in the section on copolymers, were subsequently chain extended with St to produce novel block copolymers, with an M_n>50,000 and a PDI<1.5 [102]. Although the blocking efficiency was not calculated, the gel permeation chromatography (GPC) results indicate that the macroinitiator functionality was high, as evidenced by the movement of the macroinitiator peak to higher molecular weights with the chain extension [102]. Likewise, copolymers of St with NVC were also chain extended with St to prepare block copolymers [110]. However, even though the authors claim chain-end functionality was high, the polydispersity of the final block copolymer increased from 1.21 to 1.45 and the GPC trace contained a long tail to lower molecular weight, indicating the presence of a significant amount of unextended macroinitiator [110]. Baumert and Mülhaupt used TEMPO-terminated pSt to initiate the copolymerization of St with AN. pSt prepared containing an α-carboxy functionality was used successfully to prepare the functional copolymers, which were proposed for different types of post-functionalization reactions [101].

Armes et al. prepared TEMPO-capped poly(sodium 4-styrenesulfonate) (SSt) using potassium persulfate as the initiator in refluxing solution of a 3:1 v:v mixture of ethylene glycol and water at 120 °C [150], which was then used as a macroinitiator for the successful preparation of block copolymers with 4-(dimethylamino)methylstyrene (DMAMS) and sodium 4-styrenecarboxylate (SSC). The pSSt-b-pDMAMS block copolymer formed an insoluble zwitterionic complex when exposed to an acidic environment, which was reversible upon addition of base. The formation of the pSSt-b-pSSC block copolymer was confirmed using dynamic light scattering because the block copolymers could not be analyzed using GPC [150]. When the weakly acidic carboxylate groups became fully protonated, the block became hydrophobic, resulting in micelle formation with the size of the micelles dependent on the content of pSSt in the block copolymer. Scheme 13 contains an illustration of this system.

Nowakowska et al. also used nitroxide-capped pSSt macroinitiators (in this case, HTEMPO) as the macroinitiator for the preparation of a block copolymer [151]. Their goal was to incorporate vinyl naphthalene (VN) and use the resulting block copolymers as photocatalysts. Although there was little data presented to support the conclusion that the polymerization was "living", there was significant spectroscopic evidence to confirm the presence of some pVN in the block copolymer and that the behavior of the block copolymer was different from the statistical copolymer [151].

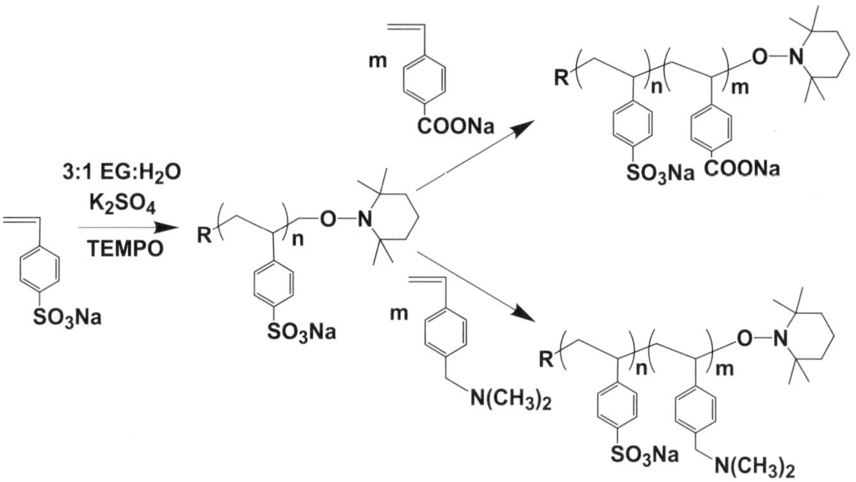

Scheme 13. Synthesis of water soluble block copolymers by TEMPO-mediated polymerizations [150]

Table 3. Summary of St-based block copolymers prepared using nitroxide-mediated CRP methods

Macroin.	Block	Nitroxide	Comments	Investigator
p4-CMSt	St	TEMPO	p(CMSt) homopolymer remained	Boutevin et al. [140, 142]
pCSt	St	MOTEMPO	Blocking efficiency function of substituent position: 2>3>4	Yoshida and Fuji [141]
pBrSt	St	MOTEMPO	Bimodal GPC, dead macroinitiator	Yoshida [99]
pSt	BrSt	MOTEMPO	Monomodal, M_w/M_n ↑ 1.14 to 1.34	Yoshida [99]
pSt	tBuSt	A-T	Lower T than TEMPO, <30% conversion of tBuSt to avoid network formation	Catala et al. [143]
ptBOSt	St	TEMPO	Controlled, hydrolysis to vinyl phenol units	Fukuda et al. [145]
pSt	PIMS	TEMPO	Deprotected to form amino-functional blocks	Laus et al. [146]
pSt	MPCS	TEMPO	Rod-coil diblock copolymer, LC behavior at 27–37 mol% MPCS	Wan et al. [147, 148]
pAcOSt	MPVB	TEMPO	Smetic-isotropic transitions observed	Ober et al. [149]
p(St-r-CMI)	St	TEMPO	High chain end functionality	Schmidt-Naake and Butz [102]
p(St-r-NVC)	St	TEMPO	M_w/M_n ↑ 1.21 to 1.45, tailing	Schmidt-Naake and Butz [102]
pSt	St/AN	TEMPO	Controlled, high blocking efficiency	Baumert and Mülhaupt [101]
pSSt	DMAMS	TEMPO	Zwitterionic complex in acid	Armes et al. [150]
pSSt	SSC	TEMPO	Micelle formation when SSC protonated	Armes et al. [150]
pSSt	VN	HTEMPO	Block behavior different from statistical copolymer	Nowakowska et al. [151]

Table 3 contains a summary of the types of St-based block copolymers prepared using nitroxide-mediated CRP methods. As detailed in the above discussion and in the table, using TEMPO as a mediator for St-based monomers results in successful sequential polymerizations to form block copolymers, even for monomers with bulky substituents such as those containing LC functionalities. Some side reactions are apparent, however, when halogenated St-based monomers are used, leading to lower blocking efficiencies. The following section describes the difficulties associated with incorporating other monomers into block copolymers using the TEMPO system and some strategies used to overcome them.

3.1.1.2
Acrylate Monomers

There have been attempts to use TEMPO-mediated polymerizations to produce block copolymers with monomers other St-based monomers. One of the earliest was by Georges et al., who used 4-oxo-TEMPO (OTEMPO) in the presence of AIBN for the polymerization of nBA, which was followed by a chain extension with St [152]. Although the authors claimed that this system was "living", they offered no monomer conversion data and the molecular weight distributions for either pnBA or pSt-*b*-pnBA were significantly higher than for pSt polymerized with the OTEMPO system (M_w/M_n=1.29–1.53 vs. 1.14) [152]. This indicates that the nBA system was less controlled than the St system. No molecular weight data was provided for the chain extension of pnBA with *t*-butyl acrylate (tBA); however, the GPC traces indicate some dead polymer chains based on a long low molecular weight tail.

Zaremski et al. investigated the polymerization of MA using TEMPO and pSt TEMPO as (macro)initiators [153]. Their initial results suggested that, due to the dynamics of the equilibrium, even a small amount of excess TEMPO in the system affected the rate of polymerization of MA and that to have a successful polymerization, the ratio of active chains and nitroxyl radicals should be as close to one-to-one as possible. Using pSt-TEMPO macroinitiators, which conform to this requirement, they found that although the pSt can act as an initiator, it was consumed slowly during the polymerization and the monomer conversion of MA did not increase above 60–70% [153]. The initial rate of polymerization decreased with time and began to fall off significantly around 60% monomer conversion. Analysis of the extracted products showed no evidence of pSt macroinitiator, nor pMA homopolymer, only the block copolymer. The authors suggested that the polymerization changed from a living mechanism to a "dead end" one as a result of a build-up of free TEMPO in the system from bimolecular termination. EPR analysis showed that free TEMPO was present, with a concentration on the order of 10^{-5} mol/l, which supported the idea that the polymerization could no longer continue [153].

Barbosa and Gomes used the AIBN/TEMPO system to homopolymerize a side-chain LC acrylate monomer, 4'-ethylbiphenyl-4-(4-propenoyloxy-butyloxy)benzoate (EBPBB, Fig. 11) [154]. The polymerization was carried out at 135 °C for 48 h, resulting in a monomer conversion of 78% and a polymer M_n=

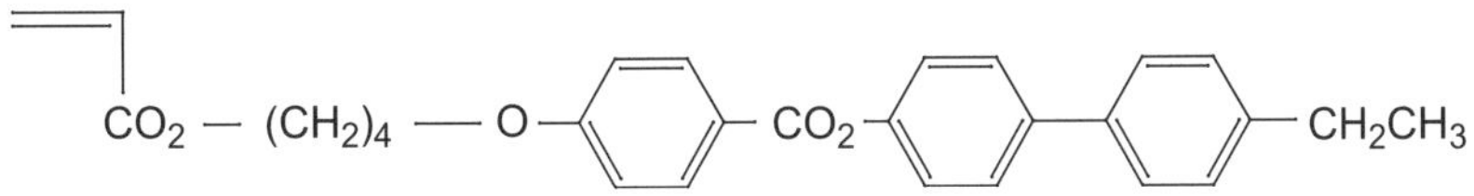

Fig. 11. 4'-Ethylbiphenyl-4-(4-propenoyloxybutyloxy)benzoate (EBPBB) [154]

6900 with $M_w/M_n=1.47$. ^{1}H NMR analysis of the product confirmed the presence of TEMPO-capped chain ends, but the amount was not quantified. Chain extension with St (16 h, 34% monomer conversion) produced a copolymer with an increased molecular weight ($M_n=9700$); however, the molecular weight distribution nearly doubled ($M_w/M_n=2.61$) [154]. Although the authors showed there was an LC phase transition centered at 141 °C and claimed the polymerization was "living", the results suggest problems similar to those discussed for the above nBA/St systems.

Gnanou et al. reported the first successful homopolymerization of nBA in 1997 [70, 72, 155] using a new nitroxide, N-tert-butyl-N-[1-diethylphosphono-(2,2-dimethylpropyl)] nitroxide (DEPN, Fig. 10). This nitroxide not only afforded faster polymerization rates at lower temperatures for the polymerization of St, but also allowed the controlled polymerization of nBA [73, 156]. Further investigation showed that the success of the system lay in changing the equilibrium between the dormant chain and active species to favor the formation of a higher concentration of active species and therefore required the addition of free DEPN for full control [73] via the persistent radical effect [56]. Gnanou et al. also demonstrated that N-tert-butyl- [1-phenyl-(2-methylpropyl)] nitroxide (BPPN, Fig. 12) could be used to moderate the polymerization of St, resulting in polymers with $M_w/M_n<1.10$ [156].

Hawker et al. prepared the 1-phenylethyl adduct of BPPN, i.e., 2,2,5-trimethyl-3-(1-phenylethoxy)-4-phenyl-3-azahexane, (TMPAH, Fig. 13) and found that it was useful for the controlled homopolymerizations of St, nBA, acrylonitrile, and N,N-dimethylacrylamide [71]. For example, the homopolymerization of DMA resulted in polymers with $M_n=4000–55,000$ with $M_w/M_n=1.15–1.21$. TMPAH was also used to prepare random copolymers containing St or nBA and the above monomers, in addition to copolymers with MMA, acrylic acid, 2-hydroxyethyl acrylate (HEA), and glycidyl acrylate. As with DEPN, it was necessary to add the free nitroxide to mediate the polymerization rate, but the resulting

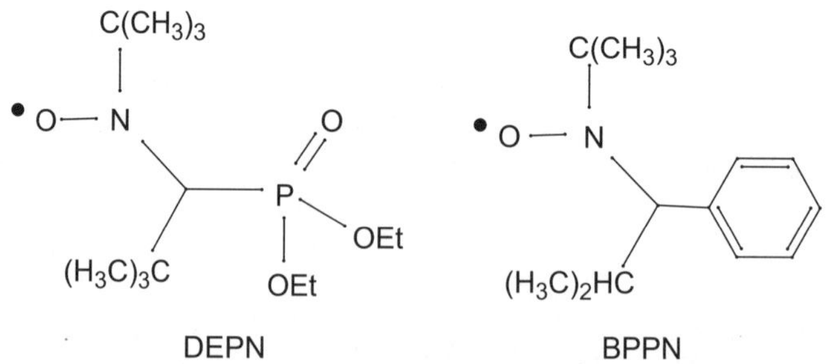

Fig. 12. Structures of N-tert-butyl-N- [1-diethylphosphono-(2,2-dimethylpropyl)] nitroxide (DEPN) and N-tert-butyl- [1-phenyl-(2-methylpropyl)] nitroxide (BPPN) [155, 156]

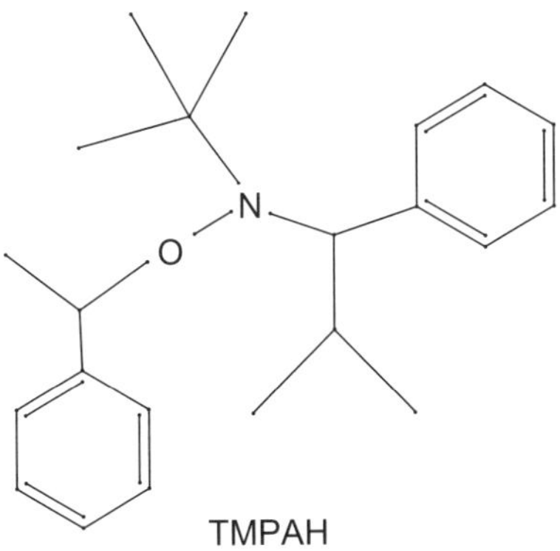

Fig. 13. 2,2,5-Trimethyl-3-(1-phenylethoxy)-4-phenyl-3-azahexane (TMPAH) [71]

polymers had predictable molecular weights and narrow molecular weight distributions, a significant improvement over the TEMPO system [71]. Block copolymers of nBA and St were possible, but only when low molecular weight pSt macroinitiators were used and a high degree of polymerization was targeted for the nBA block. Proceeding in the reverse order, however, using a pnBA macroinitiator, the polymerization was successful, producing well-defined pSt blocks with narrow molecular weight distribution ($M_w/M_n=1.19$) [71].

3.1.1.3
Diene Monomers

Georges et al. have reported on the use of TEMPO-capped pSt to prepare block copolymers with dienes like isoprene (IP) and 1,3-butadiene (BD) [157]. The authors prepared low molecular weight pSt, characterized it by ^{1}H NMR, then chain extended it with BD. They showed that the resonances associated with the pSt-TEMPO chain end disappeared and were replaced by the resonances associated with a pBD-TEMPO chain end. There was a shift in the GPC trace to higher molecular weights and a decrease in the molecular weight distribution upon chain extension. Subsequently, pSt macroinitiators were chain extended with IP, followed again by St polymerization [157]. This produced an unsymmetrical ABA triblock copolymer; however, with the third monomer addition, the molecular weight distribution was fairly unsymmetrical, although the molecular weight distribution decreased slightly ($M_w/M_n=1.30$ vs 1.24). The increase in the molecular weight was not large enough to access the chain end functionality of the pSt-b-pIP copolymer macroinitiator, but the lack of shift of the entire trace to higher molecular weights may indicate a limited chain end functionality [157].

Later work focused on preparing homopolymers of nBA and IP in the presence of an added reducing agent (either glucose or α-hydroxy ketones) [158].

This resulted in some improvement of the polymerizations due to a decrease in the concentration of free TEMPO and allowed for an increased reaction yield, but only the IP polymerization appeared to be well-controlled [158]. The polydispersities of the pnBA prepared in the presence of either glucose or hydroxyacetone were always >1.5, with significant tailing to lower molecular weights in the GPC traces.

TMPAH was used successfully for the homo-, co-, and block polymerizations of IP. In this case, due to the less reactive diene monomer, no additional free nitroxide was necessary to control the polymerization and both low and high molecular weight polymers (M_n=4500 to M_n=100,000) with narrow molecular weight distributions (M_w/M_n=1.07–1.3) were synthesized [159]. Copolymers with various styrene and (meth)acrylate derivatives, including acrylic acid and HEMA, were obtained, with the content of isoprene varying from 10% to 90% in the comonomer feed. Block copolymers were also produced, starting from either ptBA or pSt macroinitiators; however, the alternate order of blocks (i.e., starting from a pIP macroinitiator) was only achieved with St. Chain extension with tBA resulted in inefficient initiation [159], as had been found for pSt-pnBA block copolymers [71].

3.1.1.4
Vinylpyridine

Jaeger et al. homopolymerized 4-VP using the TEMPO system; however, at higher monomer conversions, the experimental molecular weights deviated from linearity, indicating that chain end functionality was lost [160]. The authors used the p4VP as a macroinitiator for chain extension with St and, although the resulting block copolymers showed an increase in molecular weight, there was tailing to lower molecular weights confirming that chain end functionality was indeed lost during the macroinitiator preparation. Subsequent quaternization of the homopolymers of the 4VP with methyl iodide, betaine formation via bromoesterification and hydrolysis of the ester, or N-alkylation with 1,3-propanesultone produced ionically charged water-soluble polymers, which was the ultimate goal of the work (Scheme 14) [160].

^{1}H NMR analysis indicated that 100% of the groups were transformed with the first two methods; however, the degree of alkylation only reached 84% conversion [160]. The authors continued to try to produce ionically charged polymers by synthesizing block copolymers of CMSt and St, followed by reacting the chloromethyl groups with trimethyl amine to produce cationic amphiphilic block copolymers capable of micelle formation [161]. Although the block copolymers had narrow molecular weight distributions, they showed evidence of unreacted macroinitiator in the GPC traces, similar to the results found by Boutevin et al. [142]. Nevertheless, the authors showed that as the length of the hydrophobic block (pSt) increases, the molecular weight of the micelles increases, as

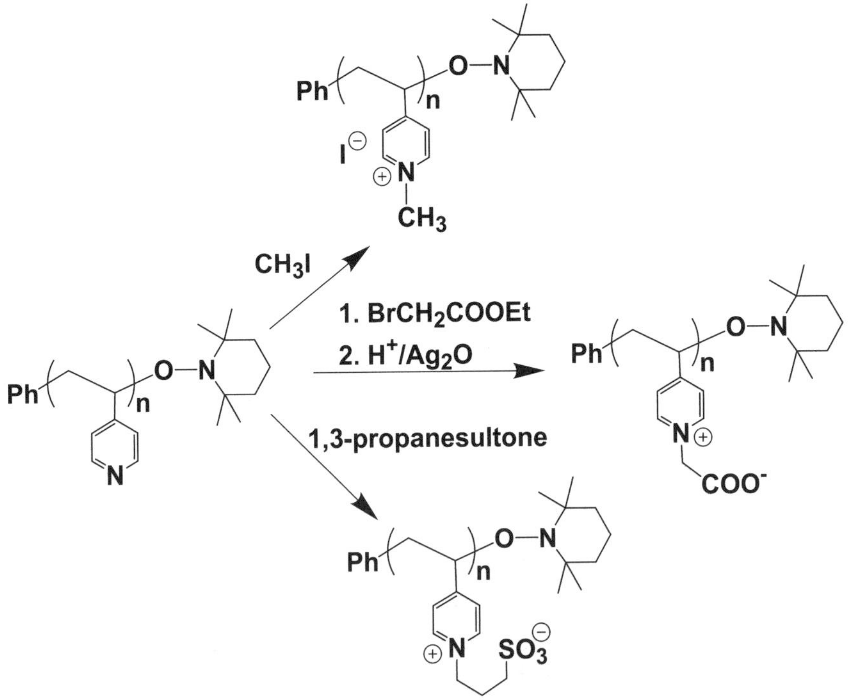

Scheme 14. Methods to modify p4VP homopolymers to produce cationic polyelectrolytes [160]

does the aggregation number, thereby providing a route to alter the micellar properties [161].

3.1.1.5
N,N-Dimethylacrylamide

Li and Brittain reported on the TEMPO-mediated polymerization of another water-soluble monomer, *N,N*-dimethylacrylamide (DMA) [162]. They found that a ratio of TEMPO:AIBN of 1 and a temperature of 120 °C provided polymers with the relatively modest molecular weight distributions (M_n=10,600, M_w/M_n=1.55). Altering the polymerization conditions to try to obtain higher molecular weight polymers was not successful. Chain extension of a pSt-TEMPO macroinitiator with sufficient DMA to achieve a molecular weight of 20,000 for the second block resulted in a limited conversion of DMA and little formation of a block copolymer [162]. The polymerization was not "living" and these results are not unexpected (see below), due to the acrylate-like structure of the DMA [152].

3.1.1.6
Maleic Anhydride

Benoit et al. reported the preparation of alternating copolymers of St with MAh using TMPAH, which was followed by a successful chain extension with St. Although no absolute molecular weight data was given, the GPC trace of the alternating-block copolymer showed little tailing to lower molecular weight. Further characterization by DSC showed the T_g of the pSt block at 105 °C and the alternating portion at 155 °C [104].

3.1.1.7
Methacrylates

Although early attempts were made to incorporate methacrylic-based monomers into copolymers using TEMPO-mediated reactions [109, 163] the controlled synthesis of homopolymers, or even block copolymers of methacrylate-based monomers, has yet to be achieved using a nitroxide-mediated polymerization. Müllen et al. attempted to prepare a homopolymer of MMA using TEMPO; however, they found that the polymerization had limited monomer conversion and very broad molecular weight distributions (M_w/M_n>1.80) [163]. When chain extension of a pSt-TEMPO macroinitiator with MMA was attempted, no increase in the molecular weight was observed. In the presence of camphorsulfonic acid the molecular weight increased; however, the molecular weight distributions were bimodal. The authors attributed this to irreversible termination of the pSt macroinitiators either during the original polymerization or from a proton abstraction once cross-propagation to MMA took place [163]. Lokaj et al. reported on the synthesis of p(St)-p(N,N-(dimethylamino)ethyl methacrylate) (DMAEMA) by TEMPO-mediated polymerization [164]. DMAEMA is a water soluble monomer and produces amphiphilic block copolymers. However, they found that the monomer conversion was limited and did not increase with increased reaction times [164]. Work by Vairon et al. proved unequivocally that the majority of the pnBMA chain ends were unsaturated resulting from β-hydrogen abstraction by the TEMPO [165]. The authors suggested that homopolymerization of methacrylates using the TEMPO system was not possible because the ratio of the rate constant of decomposition of the chain end to the rate constant of recombination prevented a controlled reaction [165]. Block copolymers were possible, however, when pSt-TEMPO was used as the macroinitiator, but although block copolymers formed, MALDI-TOF MS analysis indicated terminal unsaturation, as in the homopolymerization [165]. The only way to overcome this deficiency will be to alter the structure of the nitroxide to such an extent that chain end functionality is maintained.

3.1.1.8
Summary

Nitroxides have been used to prepare numerous block copolymers. Initially, using the TEMPO moiety, only styrene-based monomers could be incorporated into copolymers, but with the use of new nitroxides like DEPN and BPPN, the list has expanded to include acrylate-type monomers, as well as dienes, something that previously could only be accomplished through ionic mechanisms. Unfortunately, chain extension of either St or diene-based macroinitiator with an acry-

Table 4. Summary of block copolymers prepared using nitroxide-mediated polymerizations that contain one non-styryl-based block

Macroin.	Block	Nitroxide	Comments	Investigator
pnBA	St	OTEMPO	$M_w/M_n \uparrow$ 1.14 to 1.26–1.53	Georges et al. [152]
pnBA	tBA	OTEMPO	Tailing to indicate dead chains	Georges et al. [152]
pSt	MA	TEMPO	"Dead-end" polymerization due to excess free TEMPO	Zaremski et al. [153]
p4VP	St	TEMPO	Tailing indicating dead chains; modification to ionically charged blocks	Jaeger et al. [160]
pCMSt	St	TEMPO	Unreacted macroinitiator; produced micelles by reaction with trimethyl amine	Jaeger et al. [160]
pSt	DMA	TEMPO	Limited conversion of DMA	Li and Brittain [162]
pEBPBB	St	TEMPO	$M_w/M_n \uparrow$ 1.47 to 2.61; LC phase transition observed	Barbosa and Gomes [154]
pSt	BD	TEMPO	$M_n \uparrow$, $M_w/M_n \downarrow$	Georges et al. [157]
pSt	IP, then St	TEMPO	$M_n \uparrow$, $M_w/M_n \downarrow$ with both additions, but traces unsymmetrical	Georges et al. [157]
pSt	nBA	TMPAH	Must use pSt with low M_n, large ratio of pSt:nBA	Hawker et al. [71]
pnBA	St	TMPAH	Successful, narrow M_w/M_n	Hawker et al. [71]
ptBA	IP	TMPAH	Composition from 20–90% IP in copolymers, M_w/M_n<1.25	Hawker et al. [159]
pSt	IP	TMPAH	Composition from 20–90% IP in copolymers, M_w/M_n<1.25	Hawker et al. [159]
pIP	St	TMPAH	Composition from 20–60% IP in copolymers, M_w/M_n<1.30	Hawker et al. [159]
pIP	tBA	TMPAH	Inefficient initiation, unreacted pIP	Hawker et al. [159]
p(St-*alt*-MAh)	St	TMPAH	T_g pSt=105°C; alternating portion T_g=155°C	Hawker et al. [104]
pSt	MMA	TEMPO	No increase in M_n	Müllen et al. [163]
pSt	DMAEMA	TEMPO	Limited conversion of DMAEMA	Lokaj et al. [164]
pSt	nBMA	TEMPO	Unsaturated chain ends	Vairon et al. [165]

late monomer results in low blocking efficiency, potentially limiting the usefulness of this technique for the preparation of certain types of materials, i.e., thermoplastic elastomers. Methacrylate-based monomers can only be incorporated into the polymers via copolymerization; controlled homopolymerizations are not possible, most likely due to the imbalance between the rate constants of recombination and decomposition of the chain end. Success in polymerizing the methacrylate monomers may require further structural variations of the nitroxides. Table 4 summarizes the results presented in above sections.

3.1.2
ATRP

In order to highlight the comparison between nitroxide-mediated controlled radical polymerizations and ATRP reactions, the ATRP section begins with a discussion of methacrylate-based block copolymers. This is followed by copolymers of methacrylates with other monomers like acrylates, styrenes, and vinyl pyridine. A section detailing different styrene/acrylate combinations follows. Historically, these copolymers were the first literature example of well-defined block copolymers prepared using any CRP technique [42], but this has not garnered as much interest as the methacrylate-based block copolymers, as evidenced by the significantly smaller number of literature reports on these systems. Following this, block copolymers with hydrophilic and fluorinated segments are summarized. Together these sections demonstrate the adaptability of using the ATRP process to achieve diverse goals.

3.1.2.1
Methacrylates

ATRP has never suffered from the drawbacks associated with the nitroxide systems for incorporating the methacrylate monomers into copolymers. Well-defined methacrylate block copolymers have been successfully prepared using several ATRP catalytic systems. Jerome et al. used the homogeneous bis (orthochelated) Ni(II) system [166] while Sawamoto et al. used $RuCl_2(PPh_3)_3$ in the presence of $Al(OiPr)_3$ [167], both for the synthesis of wholly methacrylate block copolymers. The copolymers had predictable molecular weights and narrow molecular weight distributions [166, 167]. Amphiphilic block copolymers containing methacrylates have also been prepared [168, 169], in addition to copolymers containing fluorinated blocks [170]. They were prepared using Cu-mediated ATRP with bpy or branched/linear amines as ligands [171–174]. Details can be found in the corresponding sections of the review and the structures of the corresponding catalysts are shown in Fig. 14.

There is one difficulty common to all the CRP systems, however, and that is the adjustment of the reactivity of the end group vs that of the monomer. While

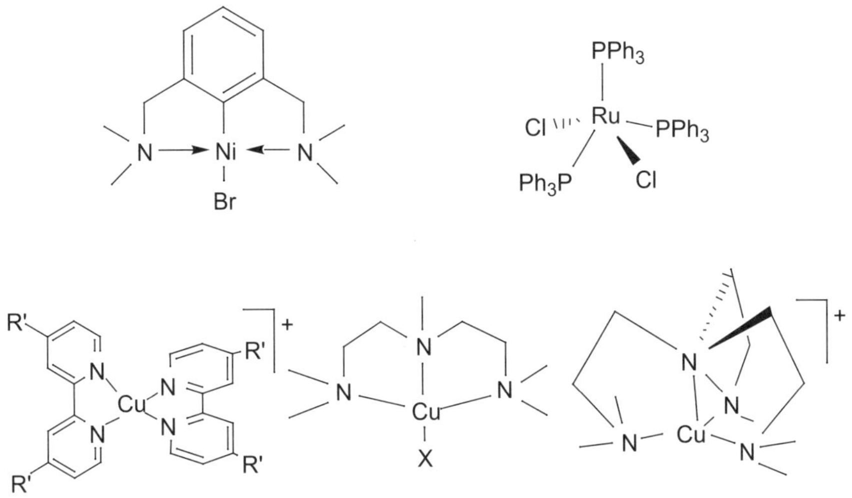

Fig. 14. Metal catalysts based on Ni, Ru, and Cu used to mediate ATRP

not as extreme as in the case of ionic systems, blocking efficiency is still affected by the rate of cross-propagation compared to the rate of polymerization of the second monomer (i.e., the rate of propagation and the equilibrium constant). This is particularly obvious when trying to polymerize methacrylate-based monomers using either pSt or polyacrylate macroinitiators because cross-propagation is slow compared to the rate of polymerization of methacrylate monomers, resulting in a low initiation efficiency and ill-defined copolymers. This sequence dependency has been overcome in the ATRP realm through the development of the halogen exchange technique [175].

3.1.2.2
Acrylates/Methacrylates

Matyjaszewski et al. introduced the concept of halogen exchange for the Cu-based ATRP reactions to combat the crossover problem [175, 176]. This technique utilizes a mixed-halogen system, i.e., a bromo-containing initiator and a Cu(I)Cl catalyst, to afford better control over initiation and polymerization of methacrylates, which is particularly useful for preparing acrylate/methacrylate block copolymers [91, 175]. The halogen exchange provides a method for adjusting the equilibrium when going from a less reactive macroinitiator (A) with an apparently smaller equilibrium constant to initiate the polymerization of a monomer that has a larger equilibrium constant (B in Fig. 15).

Neither the RAFT system nor the nitroxide system can be adjusted in this manner. When difunctional pnBA was used to prepare ABA block copolymers

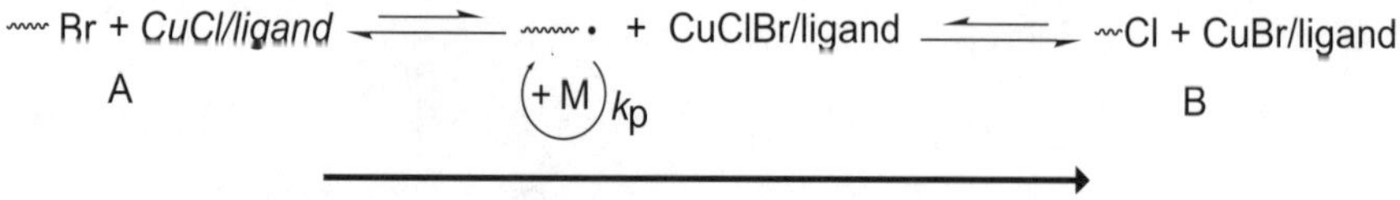

~~Br + CuCl/ligand ⇌ ~~• + CuClBr/ligand ⇌ ~~Cl + CuBr/ligand
A   (+M)k_p   B

Fig. 15. The equilibrium that occurs when the halogen exchange is applied in ATRP [175]

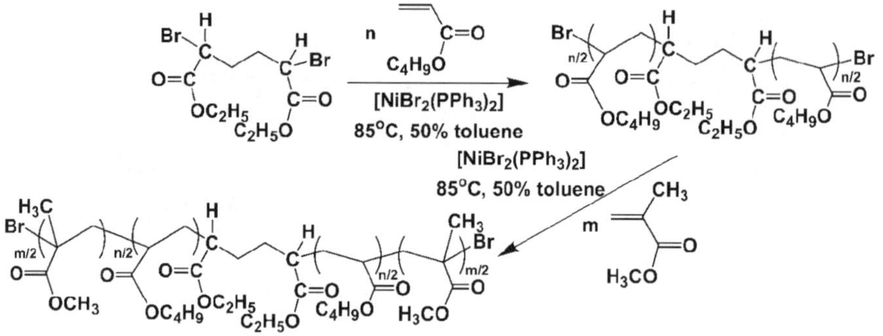

Scheme 15. Methodology for preparation of block copolymers of nBA and MMA using the NiBr$_2$ [PPh$_3$]$_2$ catalyst system

with MMA using the Ni system, Jerome et al. found that the mechanical properties of the resulting copolymers were poor compared to those of the copolymers prepared by ionic methods, presumably due to the broad polydispersity of the outer pMMA blocks (Scheme 15) [177]. Complete incorporation of the macroinitiator did not occur until >30% monomer conversion. The block copolymers did, however, microphase separate into ordered structures [178]. Matyjaszewski et al. explored the Cu-based ATRP system to prepare similar blocks. The weight and number distributions of the attempted block copolymers without using the halogen exchange are shown in Fig. 16. The "small" tailing in GPC traces observed in the RI signal, corresponds to ~50% of remaining homopolymer of pMA.

In comparison, block copolymers prepared using a copper catalyst and the halogen exchange technique had predictable molecular weights and narrower molecular weight distributions (cf. Fig. 17) [91]. There was no evidence of slow initiation and the outer pMMA blocks were more uniform. When Moineau et al. used this system, the mechanical properties were greatly improved relative to the copolymers prepared with the Ni catalyst [179]. It has also been reported that the proper choice of solvent also improves block copolymerization [180].

AB and ABA type acrylate/methacrylate block copolymers can also be prepared by ATRP using a sequential addition of monomers technique, resulting in a well-defined central block and an outer block composed of a mixture of the

3 Linear Block Copolymers

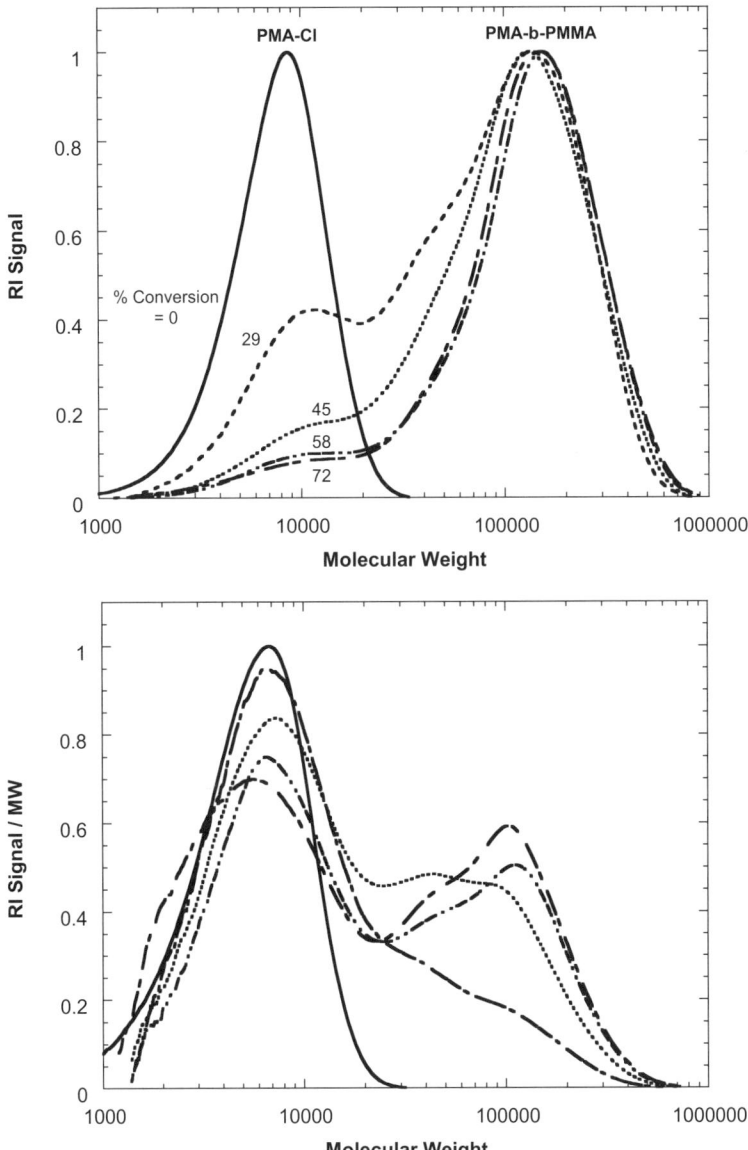

Fig. 16. Weight (top) and number (bottom) distributions of a pMA macroinitiator and a pMA-b-pMMA diblock copolymer as a function of MMA conversion when not using halogen exchange. pMA (8.3 mmol/l, M_n=6,060, M_w/M_n=1.36), MMA (5.0M), CuCl (16.6 mmol/l) dNbpy (33.2 mmol/l) in diphenylether at 90 °C. After 4.5 h (72% conversion), M_n=41,400 and M_w/M_n=3.63. Reprinted with permission from [94]. Copyright (2000) John Wiley & Sons, Inc.

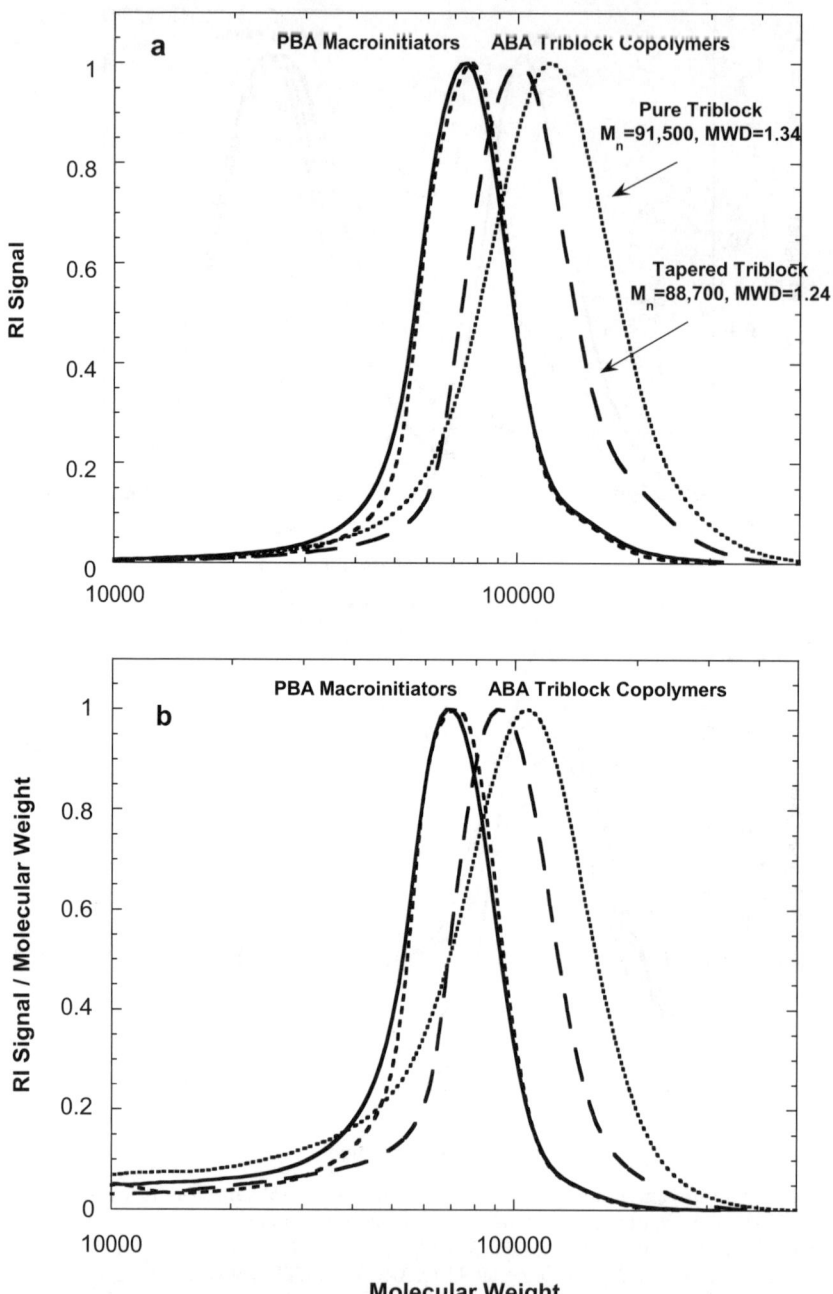

Fig. 17.a GPC traces. **b** Number distributions of high molecular weight pnBA and triblock copolymers formed during two-step and sequential addition ATRP. Reprinted with permission from [94]. Copyright (2000) John Wiley & Sons, Inc.

two monomers, as shown by Sawamoto et al. [127] and Matyjaszewski et al. [94]. Sawamoto et al. used the $NiBr_2(Pn\text{-}Bu_3)_2$ catalyst system to polymerize MMA to >90% monomer conversion (M_n=12,000, M_w/M_n=1.21), then chain extended it with both nBA (M_n=26,000, M_w/M_n=1.51) and MA (M_n=24,000, M_w/M_n=1.47) [127]. The M_w/M_n were increased upon addition of the second block, suggesting some loss of control over the polymerization.

In contrast, the copolymers prepared by Matyjaszewski et al., using halogen exchange and composed of pnBA inner segments and p(MMA-co-nBA) outer blocks, had well-defined molecular weights and narrow molecular weight distributions (M_w/M_n=1.26) [94]. Similar improvement of block copolymerization efficiency and properties of the copolymers prepared utilizing the halogen exchange was reported by Jerome et al. [179].

Another approach is to add MMA to the growing pBA chains before complete BA consumption to form gradient segments. The properties of these triblocks were different from the "clean" ABA blocks, with lowered glass transition temperatures for the mostly hard outer blocks as well as increased tensile strength and elongation, as shown in Fig. 18 [94].

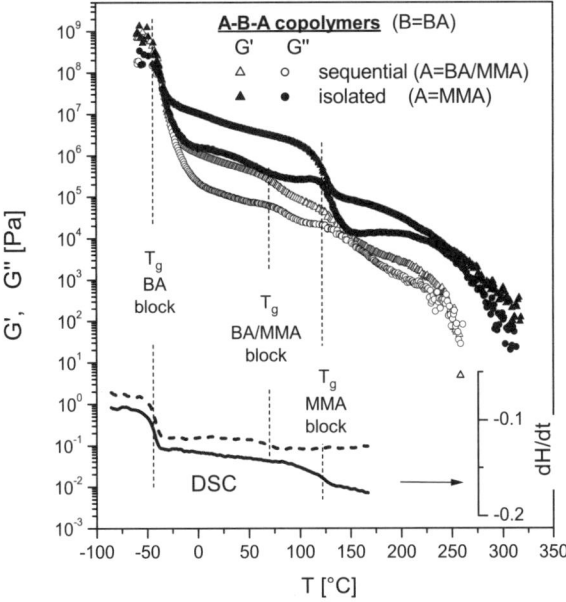

Fig. 18. Temperature dependencies of the real (G′) and imaginary (G″) component of the shear modulus measured at the deformation frequency of 10 rad/s for the pure and tapered triblock copolymers pMMA-b-pBA-b-pMMA and p(MMA-grad-BA)-b-pBA-b- p(MMA-grad-BA) of approximately the same overall composition, MW and polydispersity; DSC traces are shown to help localize the glass transition temperatures (T_g) of the microphases. Reprinted with permission from [94]. Copyright (2000) John Wiley & Sons, Inc.

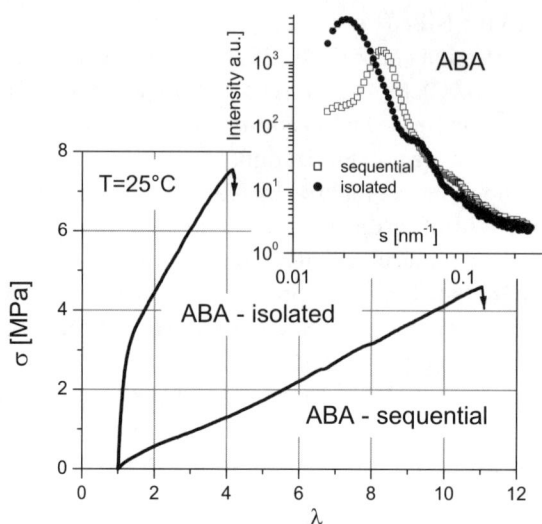

Fig. 19. The stress-strain curves recorded for the two triblock copolymer samples during cold drawing of films with a constant rate of 1 mm/min. ABA-isolated: clean pnBA central block M_n=65,200; pMMA outer blocks, M_n=13,150, overall M_w/M_n=1.34. ABA-sequential: clean pnBA central block, M_n=67,500; pMMA-grad-pnBA outer blocks: 13 mol% nBA and 87 mol% MMA, M_n=10,600, overall M_w/M_n=1.24. Inset: small angle X-ray scattering intensities for these samples. Reprinted with permission from [94]. Copyright (2000) John Wiley & Sons, Inc.

The stress-strain curves recorded for the two triblock copolymer samples during cold drawing of films with a constant rate of 1 mm/min are shown in Fig. 19. The ABA-isolated sample was a clean triblock, with a pnBA central block of M_n=65,200, two blocks of MMA with M_n=13,150, and an overall M_w/M_n=1.34. The ABA-sequential copolymer contained a clean central block of nBA of M_n=67,500, and two gradient end block copolymers containing 13 mol% nBA and 87 mol% MMA, with M_n=10,600, and an overall M_w/M_n= 1.24 [94]. The insert shows the small angle X-ray scattering intensities for these samples and correspond to cylindrical morphology of pMMA hard segments. Since both types of block copolymers had similar compositions and only varied in the sequence distribution of the monomers, this type of structural control provides yet another route to tailor make polymers with specific properties by CRP methods.

Hybrid organic-inorganic acrylate/methacrylate block copolymers have also been prepared using ATRP [181]. A pnBA bromine-terminated difunctional macroinitiator was used for the ATRP of a methacrylate monomer containing a polyhedral silsesquioxane pendant group (MA-POSS), leading to ABA diblock copolymers containing an inorganic segment [182]. Pyun and Matyjaszewski

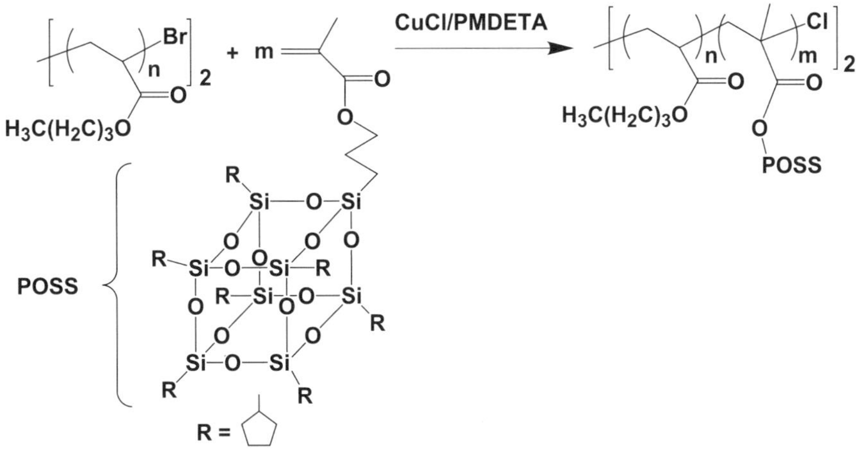

Scheme 16. Preparation of pnBA-b-p(MA-POSS) using ATRP [182]

used the CuCl/N,N,N',N'',N''-pentamethyldiethylenetriamine (PMDETA) catalyst system to invoke halogen exchange and form well-defined block copolymers (Scheme 16). GPC analysis was used to confirm that the blocking efficiency was high, and ^{1}H NMR was used to determine the molar composition (85% pnBA, 15% p(MA-POSS)) and the molecular weight (M_n=36,070), which was significantly higher and closer to the theoretical value of the M_n than that provided by the GPC analysis ($M_{n,\,GPC}$=22,800, $M_{n,\,theo}$=41,510) [182, 183].

3.1.2.3
Methacrylates/Styrenes

Zou et al. reported on the sequential copolymerization of nBMA and St by ATRP [184, 185]. nBMA was polymerized first using a 1-phenylethyl chloride initiator and a CuCl/bpy catalyst. However, in contrast to Matyjaszewski's [94] previously discussed procedure, after significant conversion of the nBMA, the reaction mixture was subjected to freeze/pump/thaw cycles to remove any unreacted monomer before the addition of St, thereby producing "clean" blocks. This procedure can be likened to precipitating the macroinitiator prior to chain extension; however, since the catalyst remains in the mix, the polymerization continues once new monomer has been added. Monomer stripping prior to the formation of the second block should be readily adaptable to industrial scale equipment. The formed block copolymers had molecular weights that ranged from M_n=15,120 to 65,790, with M_w/M_n=1.41 to 1.54 [184]. With the exception of one experiment, the molecular weight distribution broadened during the synthesis of the second block, suggesting a decreased level of control compared to the process employing purification of the macroinitiator. When the order of block synthesis was reversed and pSt macroinitiators were used for nBMA polymerization, the results were similar. Again, the molecular weight distributions broadened after chain extensions [184].

Ying et al. prepared block copolymers of both MMA and HEMA with St [186]. Chain extension of pSt-Br (M_n=6400, M_w/M_n=1.31) with MMA or HEMA using the CuCl/bpy catalyst took place at 40 °C, producing block copolymers with M_n= 13,470 and 7400 and M_w/M_n=1.33 and 1.28, respectively [186]. When pMMA-Cl (M_n=15,320, M_w/M_n=1.28) was chain extended with St, the macroinitiator was consumed slowly at 110 °C, as evidenced by the presence of a low molecular weight shoulder in the GPC traces that persisted until >40% monomer conversion. Further monomer conversion resulted in a block copolymer with M_n= 43,850 with M_w/M_n=1.40 [186]. This large increase in the molecular weight distribution is most likely due to slower dynamics of exchange associated with the Cl-based ATRP of St.

A pSt synthesized by ATRP using a CuBr/bpy catalyst was chain extended with p-nitrophenyl methacrylate (NPMA), which was subsequently transformed to its hydrolysis and amine containing derivatives by Pan et al. [187]. The homopolymerization of the NPMA using the CuBr/bpy catalyst at 110 °C was poorly controlled with M_w/M_n increasing with conversion and a concurrent change in the color of the reaction mixture from brown to green, indicating oxidation of the catalyst. The authors attributed this to coordination of the Cu(II) species to the polymer chain, preventing deactivation of the growing radicals, leading to irreversible termination. However, since Matyjaszewski et al. had demonstrated that bromine-based initiators combined with CuBr catalysts lead to poorly controlled ATRP of methacrylates [176], these results would not be unexpected and may simply be due to the generation of too many radicals, rather than interference from the polymer, particularly at this temperature under bulk polymerization conditions. Chain extension of a pSt macroinitiator (M_n= 14,730, M_w/M_n=1.18) at various NPMA:initiator ratios with the CuBr/bpy catalyst produced block copolymers with M_n=28,270–44,640 (as determined by ^{1}H NMR after conversion of the pNPMA to p(N-butyl acrylamide)) and M_w/M_n=1.26–1.36 after 24 h at 90 °C [187]. The reaction was first-order in monomer and there was a linear increase of molecular weight with conversion. The GPC traces were symmetrical. Contrary to the results seen for the homopolymerization, the color of the reaction mixture remained brown throughout the polymerization of the monomers for the preparation of the second block, presumably indicating that the majority of the catalyst remained active. The authors concluded that propagation must occur in pSt proximity, since NPMA is a poor solvent for pSt. Interaction between the NPMA block and the Cu(II) species is prevented. However, the concentration of initiating sites was approximately ten times lower in this system than for the homopolymerization, which would also decrease the amount of termination and potentially create a more controlled po-

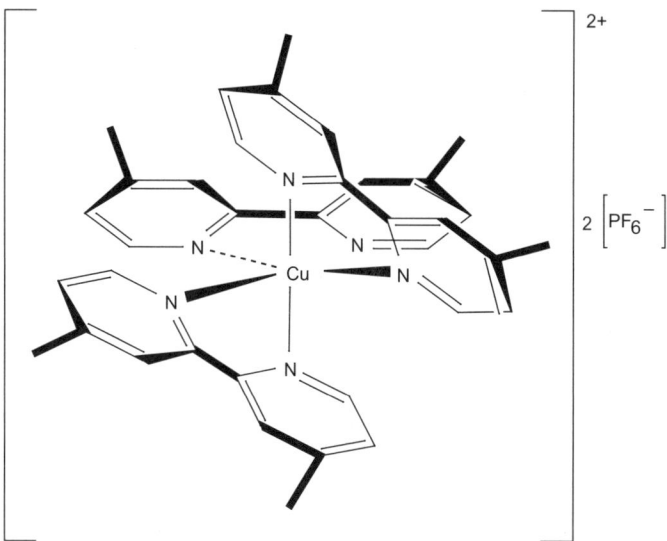

Fig. 20. Tris (4,4′-dimethyl-2,2′-bipyridine) copper(II) hexafluorophosphate [188]

lymerization. Selective solvation of these block copolymers in DMSO and CHCl$_3$ resulted in micelle formation [187].

Schubert et al. reported on the synthesis of block copolymers of pSt with MMA using a slightly different catalyst system. In the presence of Al (OiPr)$_3$, tris(4,4′-dimethyl-2,2′-bipyridine) copper(II) hexafluorophosphate can catalyze the polymerization of St, MMA, and ethyl acrylate when 1-phenylethyl bromide is used as the initiator (Fig. 20) [188]. The exact role the Lewis acid plays is still unknown, but since the Ru-based ATRP systems [43] also require the presence of a Lewis acid to be active it is assumed that in both systems the Lewis acid reduces the metal catalyst to its active low oxidation state analog. pSt-Br prepared using this Cu(II) catalyst was chain extended with MMA, resulting in a block copolymer; however, the GPC trace had a large low molecular weight shoulder, even after 27 h of polymerization, indicating a significant proportion of unreacted macroinitiator [188].

ATRP can be approached from both sides of the equilibrium, that is, beginning from an alkyl halide and a low oxidation state metal, or from a radical and the higher oxidation state metal; this latter approach is termed reverse ATRP (rATRP) [81, 189, 190]. Qiu et al. used this technique to prepare block copolymers, also of MMA and St [191]. They used a hexasubstituted ethane thermal iniferter, diethyl 2,3-dicyano-2,3-di(p-tolyl)succinate, which decomposes reversibly to form two radicals when heated. The new radical is either deactivated by the CuCl$_2$/bpy complex or adds MMA monomer, followed by deactivation, both of which will produce the dormant species in the ATRP equilibrium. The rATRP

Table 5. Summary of methacrylate containing block copolymers prepared using ATRP methods

Macroin.	Block	Catalyst	Comments	Investigator
pBMA-Br	MMA	Ni(NCN′)Br	M_n=23,200, M_w/M_n=1.15	Jerome et al. [166]
pMMA-Cl	BMA	RuCl$_2$(PPh$_3$)$_3$/ Al(OiPr)$_3$	M_w/M_n ↓ 1.26 to 1.20	Sawamoto et al. [167]
pMMA-Cl	MA	NiBr$_2$(Pn-Bu$_3$)$_2$/Al(OiPr)$_3$	M_n=24,000, M_w/M_n=1.47	Sawamoto et al. [127]
pMMA-Cl	BA	NiBr$_2$(Pn-Bu$_3$)$_2$/Al(OiPr)$_3$	M_n=26,000, M_w/M_n=1.51	Sawamoto et al. [127]
Br-pnBA-Br	MMA	NiBr$_2$(PPh$_3$)$_2$/ Al(OiPr)$_3$	M_n=90,000–156,000, M_w/M_n=1.17–1.30, slow initiation	Jerome et al. [177]
Br-pnBA-Br	MMA	CuCl/dNbpy	M_n=37,200, M_w/M_n=1.20, high blocking eff.	Matyjaszewski et al. [91]
pMMA-Cl	nBA	CuBr/dNbpy	M_n=19,000, M_w/M_n=1.15	Matyjaszewski et al. [91]
pMA-Cl	MMA	CuCl/dNbpy	M_n=41,400, M_w/M_n=3.63, ineff. initiation	Matyjaszewski et al. [91]
pMA-Br	MMA	CuCl/dNbpy	M_n=63,900, M_w/M_n=1.15, high blocking eff.	Matyjaszewski et al. [91]
Br-pnBA-Br	MA-POSS	CuCl/PMDETA	85% p(nBA), 15% MA-POSS, high blocking eff.	Pyun and Matyjaszewski [182]
pnBA-Br[a]	MMA	CuCl/HMTETA[b]	Outer block had lowered T_g than clean block, better mechanical prop.	Matyjaszewski et al. [94]
pBMA-Cl	St	CuCl/bpy	M_n=15,120–65,790, M_w/M_n=1.41–1.54	Zou et al. [184, 185]
pSt-Br	MMA	CuCl/bpy	M_n=13,470, M_w/M_n=1.33	Ying et al. [186]
pSt-Br	HEMA	CuCl/bpy	M_n=7400, M_w/M_n=1.28	Ying et al. [186]
pMMA-Cl	St	CuCl/bpy	M_n=43,850, M_w/M_n=1.40, slow consumption of macroinitiator	Ying et al. [186]
pSt-Br	NPMA	CuBr/bpy	M_n=28,270–44,640, M_w/M_n=1.26–1.36	Pan et al. [187]
pSt-Br	MMA	Cu(PF$_6$)$_2$/ dMbpy[c]	Unreacted macroinitiator	Schubert et al. [188]
pMMA-Cl	St	CuCl/bpy	Macroinitiator prepared with CuCl$_2$/bpy and iniferter	Qiu et al. [191]

[a] Sequential monomer addition without isolation of macroinitiator
[b] N,N,N',N'',N''',N'''-Hexamethyltriethylenetetraamine
[c] 4,4′-Dimethyl-2,2′-bipyridine

of MMA was controlled producing polymers with predictable molecular weights and narrow molecular weight distributions (final M_w/M_n=1.26). A pMMA-Cl macroinitiator (M_n=16,800, M_w/M_n=1.30) was used for the preparation of a block copolymer with St, forming a copolymer with M_n=113,800 and an M_w/M_n=1.59 [191]. The broadening of the molecular weight distribution, although not addressed by the authors, is most likely due to the slower dynamics of exchange for St with a chlorine-based ATRP system, but it can also be the result of an increased contribution from termination and/or transfer at the higher molecular weights.

As detailed above, various methacrylate monomers have been incorporated into block copolymers using ATRP. This method is useful for preparing wholly methacrylate block copolymers, as well as various acrylate/methacrylate and styrene/methacrylate combinations. However, halogen exchange should be utilized when a less reactive polyacrylate or pSt macroinitiator is used for the preparation of block copolymers with methacrylates to enhance the rate of cross-propagation. This will ensure complete consumption of the macroinitiator and well-defined methacrylate blocks. Table 5 contains a summary of methacrylate containing block copolymers prepared using ATRP methods. Block copolymers with hydrophilic methacrylate segments are treated separately later (cf. Sect. 3.1.2.5).

3.1.2.4
Acrylates/Styrenes

The ATRP system is not limited to just methacrylate containing block copolymers, as Wang and Matyjaszewski showed in the first example of the preparation of a block copolymer using ATRP (and in fact any CRP method) in 1995. A pMA macroinitiator was chain extended with St using the Cu(I)Cl/bpy system [42]. The copolymer with the opposite sequence of blocks was also successfully prepared [114]. Since then, there have been several reports in the literature describing the preparation of different St/acrylate block copolymer combinations using ATRP. Vairon et al. prepared block copolymers of St with nBA using conditions similar to those first reported by Wang and Matyjaszewski [120]. However, because the CuCl/bpy catalyst is heterogeneous in the non-polar reaction media, they utilized a small amount of dimethylformamide (DMF) to homogenize the system. Using 1-phenylethyl chloride (PECl) as the initiator and a reaction temperature of 130 °C, both pSt and pnBA macroinitiators were prepared. Chain extension of the pSt macroinitiator with nBA was controlled, with a constant number of propagating species during the polymerization and a linear increase in molecular weight with conversion. The alternate order of synthesizing the blocks is also possible; however, the authors indicated that conversion in the first hour of the reaction was negligible, suggesting that initiation from the chloro-terminated pnBA was slow. Subsequent chain extension of the pSt-*b*-pnBA block copolymer with more St resulted in a significant increase in the molecular weight distribution (M_w/M_n=1.56 to 1.71) and lower than predicted molecular weights ($M_{n,\,exp}$=104,300, $M_{n,\,calc}$=124,000) [120].

Matyjaszewski et al. showed, however, that when a Cu(I)Br catalyst was used in conjunction with bromine-based initiators for a similar system (St/tBA), the polymerizations were well controlled and chain extension in either direction was possible [194]. This is different from the nitroxide system where chain extension of a pSt macroinitiator with acrylate monomers resulted in low blocking efficiency [71]. AB and ABA type block copolymers were prepared using monofunctional and difunctional ptBA or pSt macroinitiators. Deprotection of the *tert*-butyl esters in these AB block copolymers afforded amphiphilic block copolymers capable of being used as surfactants for emulsion polymerizations [194, 195].

Polystyrene block was extended from pBA macroinitiator (M_n=10,200, M_w/M_n=1.22) in emulsion using Tween20 as surfactant, resulting in well-defined block copolymer (M_n=42,700, M_w/M_n=1.15) [121, 192, 193].

ABC triblock copolymers, where A, B, and C represent different monomers, have also been prepared using ATRP. Davis and Matyjaszewski chain extended a ptBA-*b*-pSt with MA, as shown in Scheme 17, confirming that the chain end functionality was high and producing an ABC triblock copolymer with an M_n= 24,800 and an M_w/M_n=1.10 [196, 197].

Other ABC triblock copolymers were prepared using mono-, di-, and trifunctional initiators to produce ABC, ABCBA, and [ABC]$_3$ linear and 3-arm star block copolymers. For example, a difunctional pSt macroinitiator (M_n=1100, M_w/M_n=1.17) was chain extended with tBA (Mn=11,200, Mw/Mn=1.16) which was subsequently chain extended with MA (M_n=27,250, M_w/M_n=1.14), as shown in Fig. 21.

A similar ptBA-pSt-ptBA macroinitiator (M_n=13,640, M_w/M_n=1.23) was chain extended with MMA, utilizing the halogen exchange technique, to produce a linear ABCBA triblock copolymer (M_n=48,470, M_w/M_n=1.21, Fig. 22) [197].

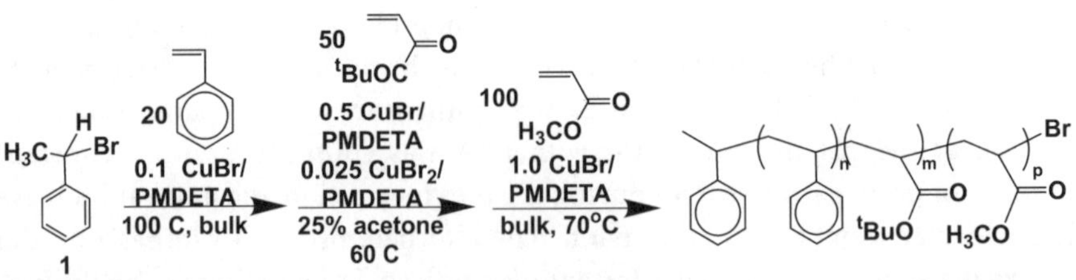

Scheme 17. Methodology for the preparation of p(St-b-tBA-b-MA) using Cu-based ATRP catalysts [196]

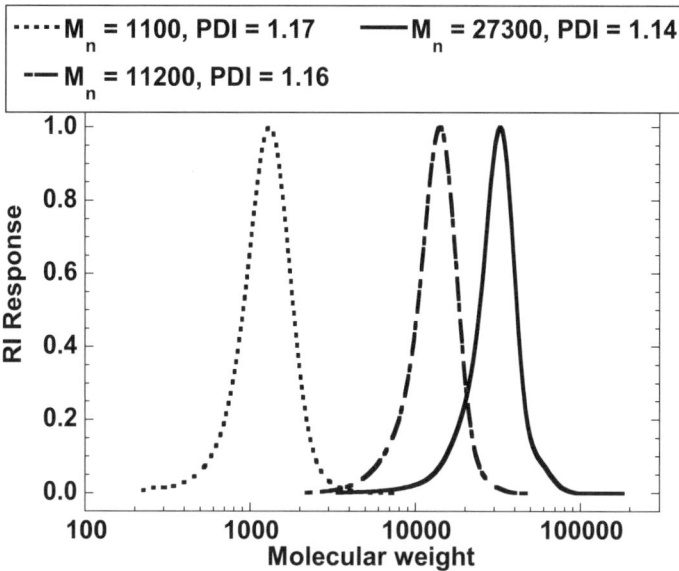

Fig. 21. GPC traces of difunctional pSt (dotted line), ptBA-b-pSt-b-ptBA (dashed line), and pMA-b-ptBA-b-pSt-b-ptBA-b-pMA (solid line). Reprinted with permission from [197]. Copyright (2001) American Chemical Society.

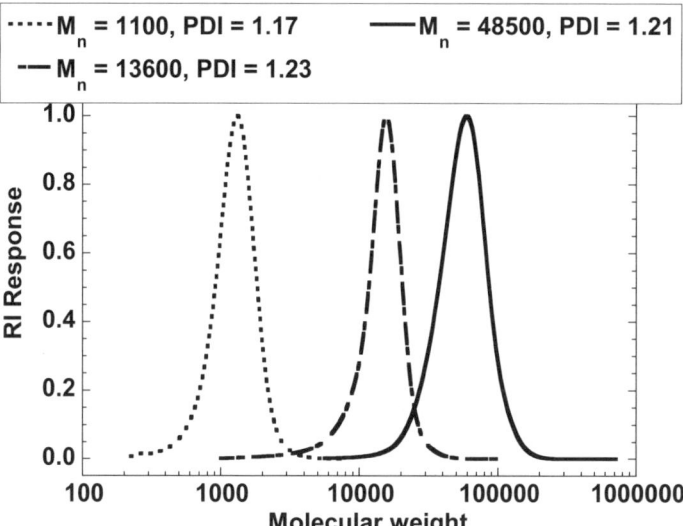

Fig. 22. GPC traces of difunctional pSt (dotted line), ptBA-b-pSt-b-ptBA (dashed line), and pMMA-b-ptBA-b-pSt-b-ptBA-b-pMMA (solid line). Reprinted with permission from [197]. Copyright (2001) American Chemical Society.

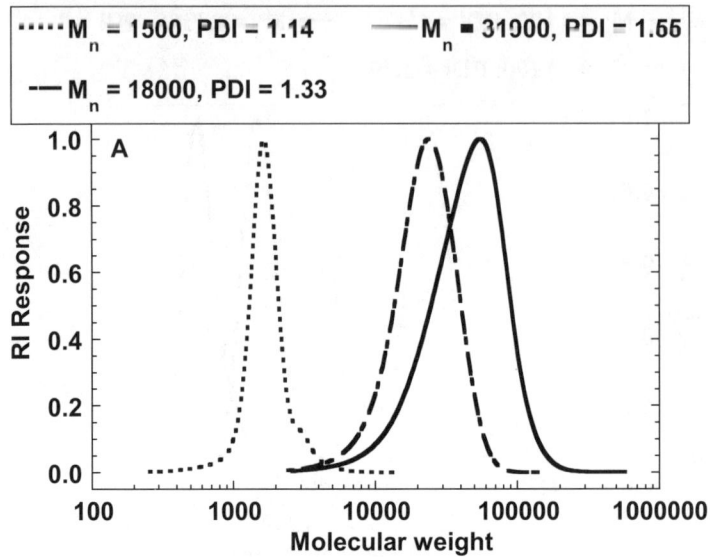

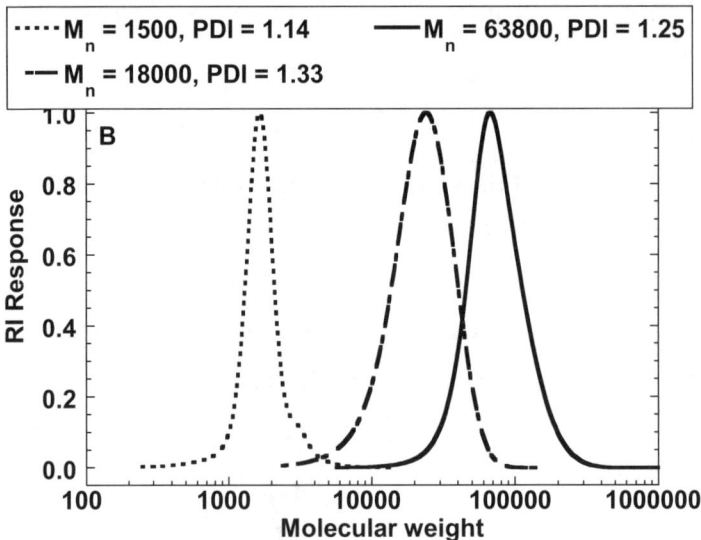

Fig. 23. GPC traces of trifunctional CH3C- [pSt-Br] 3 (dotted line), CH3C- [pSt-b-ptBA-Br]$_3$ (dashed line), and CH$_3$C-[pSt)-b-ptBA-b-pMMA-Br]$_3$ (solid line) in THF using a 1:1 molar equivalent of catalyst relative to initiator, B) GPC traces of trifunctional CH$_3$C-[pSt-Br]$_3$ (dotted line), CH$_3$C-[pSt-b-ptBA-Br]$_3$ (dashed line), and CH$_3$C-[pSt-b-ptBA-b-pMMA-Br]$_3$ (solid line) in THF using a 1:1 molar equivalent of catalyst relative to end groups. Reprinted with permission from [197]. Copyright (2001) American Chemical Society.

However, one caveat to the halogen exchange is that the ratio of end groups to the catalyst may need adjustment to obtain the best results. Figure 23 illustrates a chain extension using a 1:1 ratio of catalyst to *initiator* (A) and a 1:1 ratio of catalyst to halogen *end groups* (B). The former contained a threefold excess of chain ends relative to the concentration of catalyst and resulted in unsymmetrical chain growth. By ensuring an equimolar ratio between the chain ends and the available catalyst for exchange, growth became symmetrical and the molecular weight distribution narrowed significantly (M_w/M_n=1.25 vs 1.55) [197]. More star-like structures are described in detail in a later section.

3.1.2.5
Hydrophilic Monomers

Water soluble (meth)acrylate monomers have been incorporated into block copolymers using ATRP. Zhang and Matyjaszewski reported on the synthesis of block copolymers with DMAEMA [168]. Both mono- and difunctional halogen-terminated pMMA prepared by ATRP were used as macroinitiators for the reaction, resulting in a controlled chain extension displaying increasing molecular weights with increasing monomer conversion and narrow molecular weight distributions (M_w/M_n=1.14). Using the halogen exchange technique [175], bromine-terminated pMA macroinitiators were successfully used for the ATRP of DMAEMA in the presence of a Cu(I)Cl catalyst. Chain extension from pSt macroinitiators resulted in slow initiation and broad molecular weight distributions (M_w/M_n=1.83) [168].

Matyjaszewski et al. also prepared block copolymers of MMA with HEMA, directly producing amphiphilic block copolymers [169]. A chloro-terminated pMMA macroinitiator (M_n=3400, M_w/M_n=1.12) was chain extended with HEMA, using a 30/70 1-propanol/methyl ethyl ketone mixture as a solvent, to yield a block copolymer with M_n=32,900 with M_w/M_n=1.17, as determined by GPC analysis (Fig. 24). The synthesis of the block copolymer confirmed that the molecular weights obtained from GPC for HEMA-containing polymers were overestimates, however, since those obtained from ^{1}H NMR analysis agreed better with the theoretical values of the molecular weights calculated from the monomer conversion ($M_{n,\,NMR}$=15,000, $M_{n,\,theo}$=13,200). The GPC analysis did confirm that the chain extension was clean and blocking efficiency was high for this system [169].

Ying et al. also synthesized block copolymers of MMA and HEMA using ATRP methodologies [186]. The polymerizations also invoked the halogen exchange [175] by using an ethyl 2-bromopropionate/CuCl/bpy catalyst system and acetonitrile as a polymerization medium. The authors found that the polymerization of MMA could be controlled at 40 °C, while for the ATRP of HEMA, the temperature could be lowered to 20 °C. Chain extension of a lower molecular

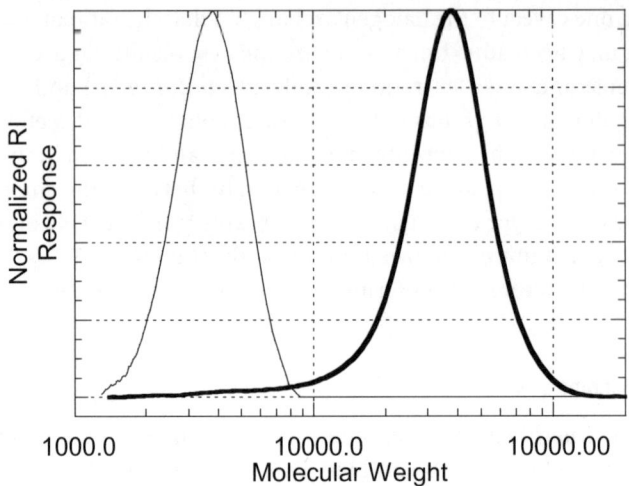

Fig. 24. GPC traces for a pMMA-Cl macroinitiator (dotted line) and pMMA-b-pHEMA (solid line). Reprinted with permission from [169]. Copyright (1999) American Chemical Society.

weight pMMA-Cl (M_n=6350, M_w/M_n=1.27) with HEMA resulted in a block copolymer with M_n=26,360 and M_w/M_n=1.54 [186]. The significant increase in the molecular weight distribution was not addressed by the authors nor were GPC traces provided to demonstrate the blocking efficiency.

Block copolymers containing VP have also been prepared by ATRP. Matyjaszewski et al. demonstrated that when the CuCl/tris[2-(dimethylamino)ethyl] amine (Me_6TREN) catalyst was used in conjunction with a chlorine-based initiator and 2-propanol as the solvent, p4VP with M_n=15,550 and M_w/M_n=1.17 could be prepared [198]. Subsequently, a pMMA-Cl macroinitiator (M_n=7700, M_w/M_n=1.07) was chain extended with VP using the CuCl/Me_6TREN system to yield a block copolymers with M_n=89,500 and an M_w/M_n=1.35, as shown in Fig. 25. There was a clean shift in the macroinitiator peak to higher molecular weights with increasing monomer conversion, indicating little end-group loss during chain extension [198]. This is in contrast to the results of the p4VP-containing block copolymers prepared using the TEMPO-mediated polymerization [160] ^{1}H NMR analysis provided M_n=62,500, in better agreement with the calculated value of M_n=63,800 based on the monomer conversion.

Matyjaszewski et al. prepared macroinitiators of nBA and trimethylsilyl-protected HEA (HEA-TMS) by ATRP using the CuBr/PMDETA catalyst system. Each of these macroinitiators was subsequently chain extended with the alternate monomer to form block copolymers using the same catalyst system [199]. AB and ABA triblock copolymers were synthesized. The sequential addition of monomers technique was used for the preparation of AB block copolymers starting with the pnBA macroinitiators, producing a small gradient of composi-

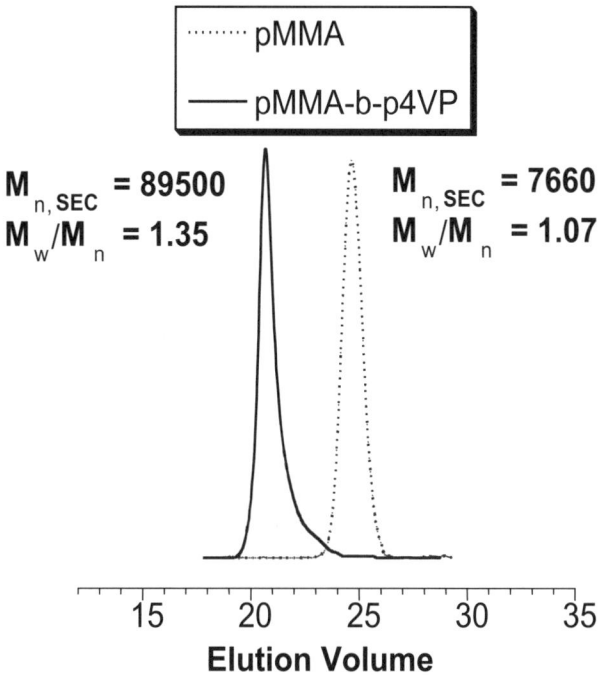

Fig. 25. GPC chromatograms of a pMMA macroinitiator and a pMMA-b-p4VP copolymer. Reaction conditions: 40 °C; $[4VP]_0$=4.62 M; $[4VP]_0/[pMMA-Cl]_0$=710; $[pMMA-Cll]_0/[Cu-Cl]_0/[Me6TREN]_0$=1/2/2. Reprinted with permission from [198]. Copyright (1999) American Chemical Society.

tion in the second block. Upon deprotection under acidic conditions, the copolymers were amphiphilic, and, in the case of the ABA block copolymers, the hydrophilic portion was either the central block or on the exterior. The composition and order of block affinity to water affected the behavior in solution [199].

In addition to attempting to use TEMPO for the polymerization of DMA, Li and Brittain tried to use ATRP to polymerize DMA, but were unsuccessful [162]. Teodorescu and Matyjaszewski, however, later reported on the ATRP of (meth)acrylamides using a different catalyst system [200]. They concluded that the problem with using "traditional" ATRP catalysts based on complexes with bpy or linear amine-based ligands was that the rate of activation was slow and the rate of deactivation was fast, limiting the conversion of the monomer. Using 1,4,8,11-tetramethyl-1,4,8,11-tetraazacyclotetradecane (Me_4cyclam) as the ligand, in conjunction with CuBr as the copper(I) species, they prepared a catalyst system which creates a "poor" deactivator. Polymers of DMA as well as N-tert-butylacrylamide and N-(2-hydroxypropyl)-methacrylamide (HPMA) were synthesized, as illustrated in Fig. 26, although the polymerizations had limiting monomer conversions, uncontrolled molecular weights ($M_{n, SEC}$=34,000) and

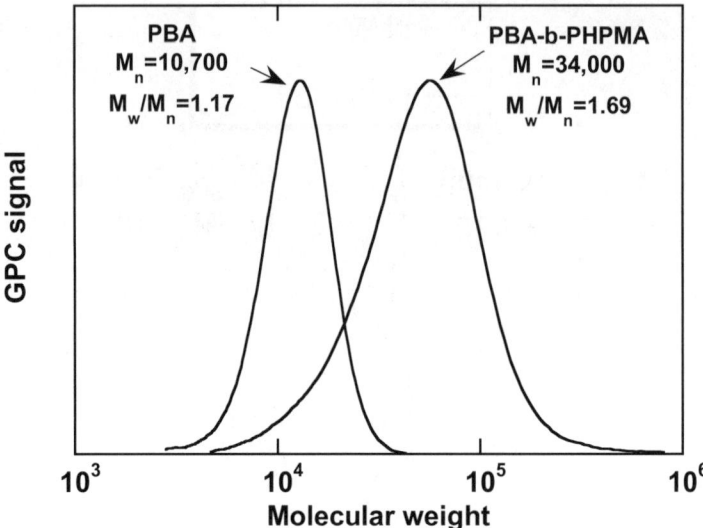

Fig. 26. Molecular weight distributions of a pMA macroinitiator and a pMA-b-pDMAA block copolymer. Exp. cond.: DMAA:pMA:CuBr:CuBr$_2$:Me$_4$Cyclam=238:1:1:0.1:1.1; solvent: methanol:ethyl acetate=1:1 (v/v); T=50°C; time=5h. Reprinted with permission from [200]. Copyright (1999) American Chemical Society.

relatively broad molecular weight distributions (M_w/M_n=1.69) [200]. The addition of Cu(II)Br$_2$, which shifts the equilibrium toward the dormant species, decreased the rate as well as the molecular weight distribution, increasing the control over the polymerization slightly. Block copolymers were prepared with well-defined pMA and pnBA macroinitiators, and although cross-propagation was efficient, the second block remained less well-defined [200].

Sawamoto et al. also attempted ATRP of DMA using the Ru-based catalyst system [201]. Although no block copolymers were reported, chain extension of the pDMA with a fresh feed of monomer resulted in an increase of the molecular weight and a decrease in the molecular weight distribution at 60 °C. However, the GPC traces were bimodal at the early stages of the polymerization, suggesting a high molecular weight portion of dead chains due to termination via coupling [201]. Nevertheless, high monomer conversions were achieved, suggesting that this catalyst system is useful for the preparation of pDMA.

Fukuda et al. used both a nitroxide-mediated polymerization as well as ATRP to polymerize a sugar containing methacrylate monomer and produce water-soluble glycopolymers; however, only ATRP was used to prepare block copolymers [202, 203]. The ATRP of 3-O-methacryloyl-1,2:5,6-di-O-isopropylidene-D-glucofuranose (MAIpGlc, Fig. 27) was carried out to determine the characteristics of the homopolymerization. It was found that the relationship between the concentration of initiator and the rate of polymerization was not a simple first-

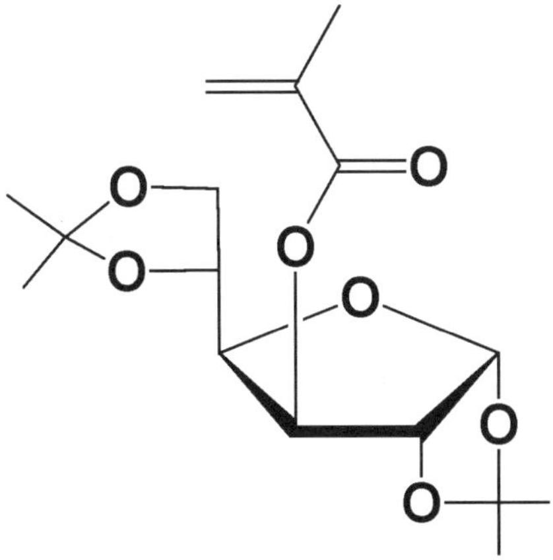

Fig. 27. 3-O-Methacryloyl-1,2:5,6-di-O-isopropylidene-d-glucofuranose (MAIGlc) [203]

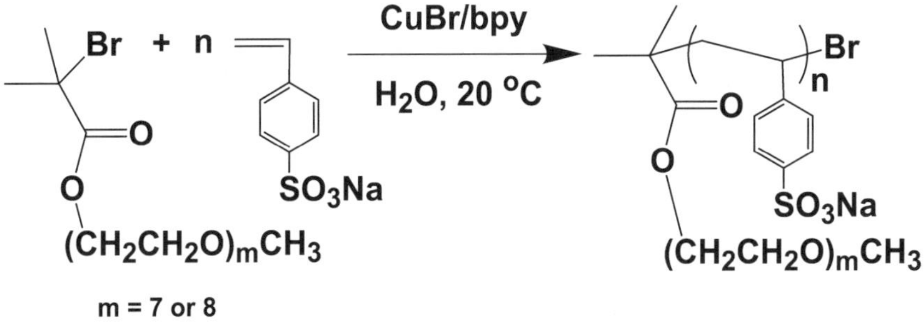

Scheme 18. Synthesis of pOEGMA-b-pNaVB copolymers using ATRP in aqueous media at ambient temperature [206]

order one, but that the molecular weights increased linearly with conversion and were predictable based on the initial degree of polymerization determined by the ratio of monomer to initiator. The resulting polymers had narrow molecular weight distributions ($M_w/M_n<1.3$). Formation of block copolymers using a bromine-terminated pSt macroinitiator was deemed successful, with an overall $M_n=14,400$, but the molecular weight distribution increased from $M_w/M_n=1.09$ to 1.34, with a significant low molecular weight tail overlapping with the macroinitiator [203]. The block copolymer was deprotected to yield an amphiphilic block copolymer and morphological characterization clearly showed pSt domains imbedded in a p(3-O-methacryloyl-1,2:5,6-D-glucofuranose) matrix [203]. Similar work was carried out by Haddleton et al., who used other natural products for initiation and side chain functionalities [204, 205].

Recent results from Armes et al. have shown that ATRP can also be used to polymerize water soluble monomers [206] at ambient temperatures, as opposed to the 120 °C needed for the TEMPO system [150]. Armes et al. polymerized oligo(ethylene oxide) methacrylate (OEGMA) using a 2-bromoisobutyrate initiator (modified with a short oligo(ethylene oxide) chain to impart water solubility) in the presence of CuBr/bpy as the catalyst at 20 °C. Chain extension of this polymer with sodium 4-vinyl benzoate (NaVB) led to the formation of block copolymers with an M_w/M_n=1.27 and contained 57 mol% pOEGMA (Scheme 18) [206]. Further study of the protonated block copolymers indicated that the styrenic block becomes hydrophobic, producing a micellar core which is surrounded by a hydrophilic pOEGMA corona.

3.1.2.6
Fluorinated Monomers

Ying et al. used bromine-terminated pMA, pnBA, and pSt mono- and difunctional macroinitiators, synthesized by ATRP, for chain extension with 2- [(perfluorononenyl)oxy] ethyl methacrylate (FNEMA) and ethylene glycol monomethacrylate mono-perfluorooctanoate (EGMAFO, Fig. 28) [207]. Chain extension with FNEMA using the CuBr/bpy catalyst system proceeded in a controlled fashion from all the macroinitiators with final M_n=12,400 to 28,660 and M_w/M_n<1.5. While the GPC underestimates the contribution of the pFNEMA block, the composition determined by ^{1}H NMR agrees with the expected values based on the amount of consumed monomer. In contrast, while the blocking efficiency was high from the acrylate-based macroinitiators when chain extended with the EGMAFO, initiation was incomplete from the pSt macroinitiators and a high molecular weight shoulder indicated coupling of chains occurred [207]. The M_w/M_n>2.0 for all chain extensions with the exception of the difunctional pMA macroinitiator, where the M_w/M_n=1.70. The composition of this block copolymer (75:25 MA:EGMAFO) agreed well with theoretical values. No attempts were made to alter the polymerization conditions to obtain more well-defined polymers. In the case of the acrylate-based macroinitiators, use of halogen exchange [175] may have allowed for better control, particularly in the second sys-

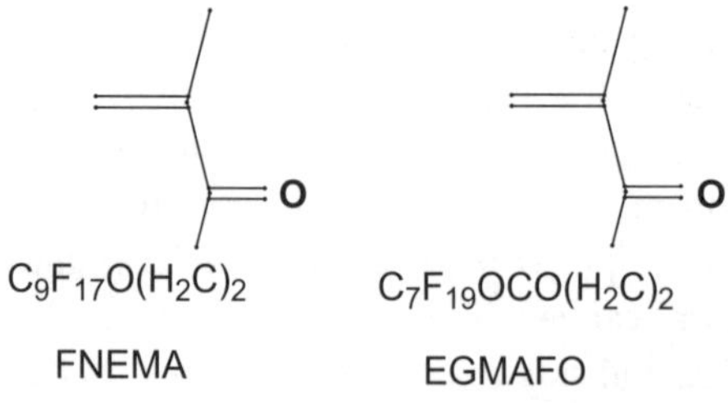

Fig. 28. Structures of 2- [(per-fluorononenyl)oxy] ethyl methacrylate (FNEMA) and ethylene glycol mono-methacrylate mono-perfluorooctanoate (EGMAFO) [207]

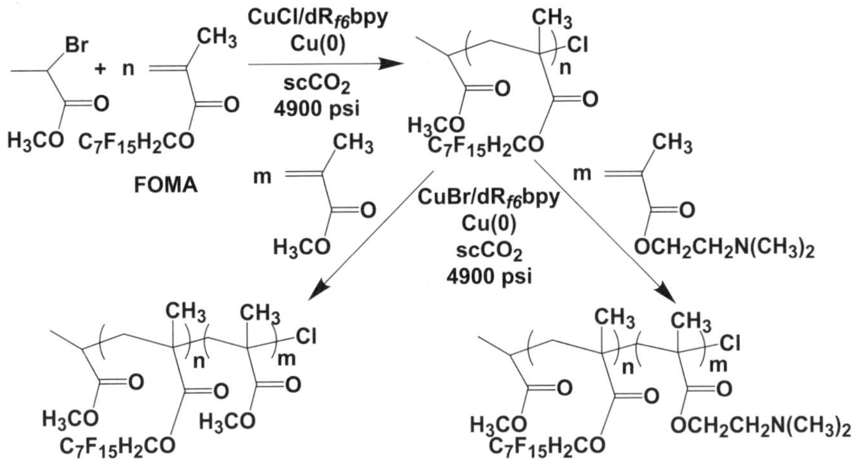

3 Linear Block Copolymers

Scheme 19. Synthesis of pFOMA-b-pMMA and pFOMA-b-pDMAEMA via ATRP in supercritical CO_2 [170]

tem. In the pSt/EGMAFO, homogenization of the catalyst system would probably decrease the amount of termination in the system.

Supercritical CO_2 (scCO_2) has also been used as a polymerization medium for the preparation of block copolymers via ATRP [170]. In another example of polymerizing fluorinated monomers, Matyjaszewski et al. used a fluoroalkyl bpy ligand, 4,4'-di(tridecafluoro-1,1, 2,2, 3,3-hexahydrononyl)-2,2'-bipyridine (dR$_{f6}$bpy), in the presence of CuCl to form a homogeneous catalyst in scCO_2. The homopolymerizations of 1,1-dihydroperfluorooctyl acrylate (FOA) and 1,1-dihydroperfluorooctyl methacrylate (FOMA) were carried out first, followed by the chain extension of the pFOMA with MMA and DMAEMA (Scheme 19). Although the block copolymers could not be characterized using GPC methods, both ^{1}H NMR characterization and DSC measurements indicated the presence of the second monomer. Further evidence for block formation came from solubility studies where the behavior of the block copolymers was changed from that of the homopolymers of the CO_2-philic monomers [170].

3.1.2.7
Block Copolymers in Dispersed Media

ATRP can also be carried out in water under homogeneous [208] or under biphasic conditions [209–211]. Block and statistical copolymers have also been prepared in water-borne systems. Matyjaszewski et al. demonstrated that a copolymerization of MMA with either BA or nBMA proceeded in a controlled fashion, resulting in copolymers with M_n=26,850 (M_w/M_n=1.22) or M_n=33,550 (M_w/M_n=1.25), respectively [121]. A pnBA macroinitiator prepared in a bulk

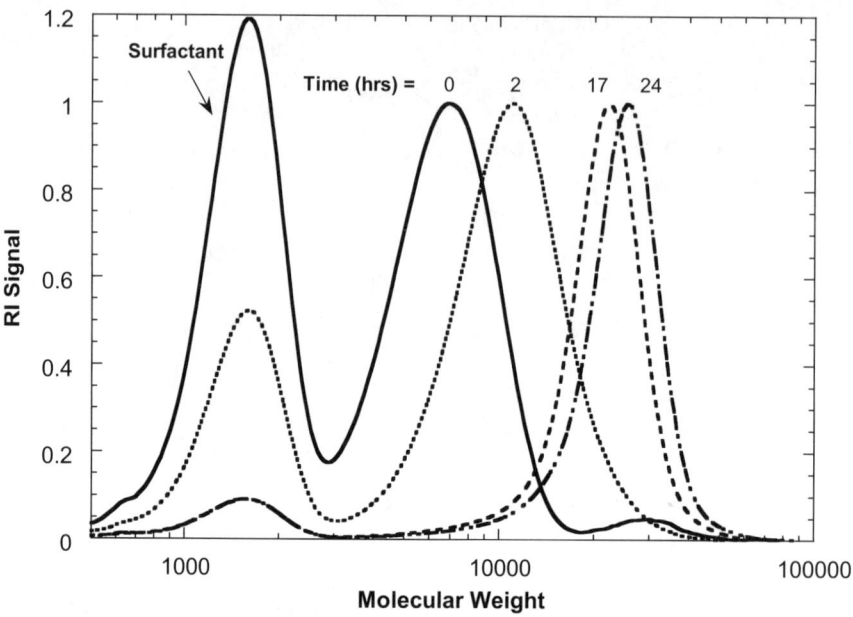

Fig. 29. Molecular weight distribution development as a function of time and monomer conversion for the water-borne chain extension of pBA by styrene with Brij98, CuBr/dAbpy, and hexadecane. Time (h)/% Conversion/M_n/M_w/M_n data: 0/0/6200/1.18, 2/25/9500/1.19, 17/70/19100/1.14, 24/78/21400/1.15. (M_n and M_w/M_n data calculated not including surfactant peak). Reprinted with permission from [121]. Copyright (2000) American Chemical Society.

ATRP polymerization (M_n=5750) was chain extended with St to yield a block copolymer with M_n=21,400 and an M_w/M_n=1.15, after 24 h and 78% monomer conversion, as shown in Fig. 29 [121]. Although the authors showed that both statistical and block copolymers could be prepared in the water-borne system, the latex stability was often low and phase separation was observed. Success in these polymerizations depends strongly on the nature of the additives present (i.e., surfactant, macroinitiator, etc.) [121, 212].

Using a fluoro-substituted linear triamine-based ligand that created a soluble catalyst in perfluoromethyl cyclohexane, Haddleton et al. demonstrated that at 90 °C the fluorous and organic MMA/toluene phases were miscible, which allowed the ATRP of MMA to occur with control [213]. Upon cooling, the phases separated, with the catalyst remaining in the fluorous phase and the polymer in the organic phase, providing a route to pure polymer without the need for further purification. The polymerization of benzyl methacrylate (BzMA) initiated by isolated pMMA was successful, producing a block copolymer with M_n= 28,900 and M_w/M_n=1.48 [213].

The success of the above polymerizations demonstrates that ATRP can be utilized to prepare a wide variety of block copolymers, ranging from those that are wholly organic soluble to those that are totally water soluble, along with amphiphiles that fall in between. Sugar and hydroxyethyl-based monomers can be incorporated to impart biocompatibility to the polymers, producing polymers that may be useful for biomedical applications. The diversity of the ligands available provides a route to tailor the polymerization catalyst to fit specific criteria, as was demonstrated by the use of fluorous ligands to create homogenous catalysts in biphasic systems. Table 6 contains a summary of block copolymers discussed in the above sections.

Table 6. Summary of styrene/acrylate, amphiphilic, and novel block copolymers prepared using ATRP methods

Macroin.	Block	Catalyst	Comments	Investigator
PSt-Cl	MA	CuCl/bpy	First ATRP block copolymer	Matyjaszewski et al. [81, 114]
pSt-Cl	nBA	CuCl/bpy	DMF as solvent, controlled chain extension	Vairon et al. [120]
pnBA-Cl	St	CuCl/bpy	DMF as solvent, slow initiation	Vairon et al. [120]
pSt-Br	tBA	CuBr/PMDETA	M_n=3725–18,300, M_w/M_n=1.14–1.44	Matyjaszewski et al. [194]
ptBA-Br	St	CuBr/PMDETA	M_n=18,520, M_w/M_n=1.15	Davis and Matyjaszewski [196]
Br-pSt-Br	tBA	CuBr/PMDETA	M_n, macro<2000, M_n, block>7000, M_w/M_n<1.5	Matyjaszewski et al. [194]
Br-ptBA-Br	St	CuBr/PMDETA	M_n=7300–21,720 M_w/M_n=1.14–1.27	Matyjaszewski et al. [194]
ptBA-b-pSt-Br	MA	CuBr/PMDETA	M_n=24,790, M_w/M_n=1.10	Davis and Matyjaszewski [196]
pMMA-Cl	DMAEMA	CuCl/HMTETA	M_w/M_n<1.20, ratio of DMAEMA:MMA=0.57 to 2.48	Zhang and Matyjaszewski [168]
Cl-pMMA-Cl	DMAEMA	CuCl/HMTETA	M_w/M_n<1.25, ratio of DMAEMA:MMA=0.62 to 1.6	Zhang and Matyjaszewski [168]
pMA-Br	DMAEMA	CuCl/HMTETA	M_w/M_n=1.15, ratio of DMAEMA:MMA=0.8	Zhang and Matyjaszewski [168]
pSt-Br	DMAEMA	CuCl/HMTETA	Slow initiation	Zhang and Matyjaszewski [168]
pMMA-Cl	HEMA	CuCl/bpy	M_n=32,900, M_w/M_n=1.17	Matyjaszewski et al. [169]
pMMA-Cl	HEMA	CuCl/bpy	M_n=26,360, M_w/M_n=1.54	Ying et al. [186]
pMMA-Cl	VP	CuCl/Me$_6$TREN	M_n=89,500, M_w/M_n=1.35, clean chain extension	Matyjaszewski et al. [198]

Table 6. contents

Macroin.	Block	Catalyst	Comments	Investigator
pnBA-Br	HEA-TMS	CuBr/PMDETA	AB and ABA blocks, % HEA=27% to 71%	Matyjaszewski et al. [199]
pHEA-TMS	nBA	CuBr/PMDETA	AB and ABA blocks, M_n=12,500–31,900, M_w/M_n=1.20–1.8	Matyjaszewski et al. [199]
pMA-Br	DMA	CuBr/Me$_4$cyclam	M_n=48,600, M_w/M_n=1.33, ill-defined DMA	Teodorescu and Matyjaszewski [200]
pnBA-Br	HPMA	CuBr/Me$_4$cyclam	M_n=18,000, M_w/M_n=1.69, ill-defined HPMA	Teodorescu and Matyjaszewski [200]
pSt-Br	MAIpGlc	CuBr/bpy	M_n=14,400, M_w/M_n=1.34, amphiphilic upon deprotection	Fukuda et al. [203]
pOEGMA	NaVB	CuBr/bpy	Aqueous polymerization M_w/M_n=1.27, 57 mol % OEGMA	Armes et al. [150]
pFOMA	MMA	CuCl/dR$_{f6}$bpy	Poly. in scCO$_2$, 61% p(MMA)	Matyjaszewski et al. [170]
pFOMA	DMAEMA	CuCl/dR$_{f6}$bpy	Poly. in scCO$_2$, 31% p(MMA)	Matyjaszewski et al. [170]
pMMA	BzMA	CuBr/fluoro triamine	Fluorous biphasic system, M_n=28,900, M_w/M_n=1.48	Haddleton et al. [213]
EBriB[a]	MMA/BA	CuBr/dAbpy[b]	M_n=26,850, M_w/M_n=1.22, water-borne	Matyjaszewski et al. [121]
EBriB	MMA/nBMA	CuBr/dAbpy[b]	M_n=33,550, M_w/M_n=1.25, water-borne	Matyjaszewski et al. [121]
p(nBA)-Br	St	CuBr/dAbpy[b]	M_n=21,400, M_w/M_n=1.15, water-borne, clean chain extension	Matyjaszewski et al. [121]

[a]Ethyl 2-bromoisobutyrate
[b]4,4'-Di(5-alkyl)-2,2'-bipyridine

3.1.3
Degenerative Transfer Processes

Several different types of block copolymers have been synthesized using RAFT methodologies. A difunctional ABA block copolymer of pnBMA and pMMA with M_n=112,200 and M_w/M_n=1.14 was prepared using a dithiobenzoate-based transfer agent [53]. Moad et al. [53] and Mayadunne et al. [54, 85] showed the versatility of RAFT by combining MMA or nBMA with St, MA with ethyl acrylate [53], and St with nBA, for block copolymers in an ABA fashion using a new transfer agent based on a trithiocarbonate structure (Fig.30) [85]. They illustrat-

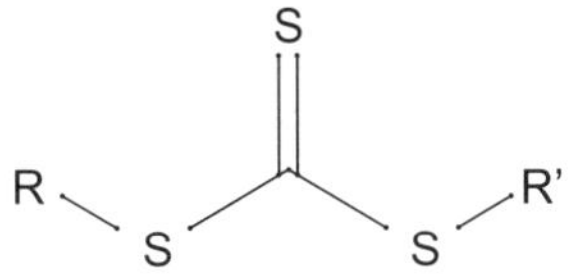

$$\underset{R\diagdown S}{\overset{S}{\underset{\|}{C}}}\diagup S\diagdown R'$$

1. R = CH$_3$, R' = C(CH$_3$)$_2$CN
2. R = CH$_3$, R' = CH(Ph)COOH
3. R = R' = CH$_2$Ph
4. R = R' = CH(CH$_3$)Ph

Fig. 30. Structure of various trithiocarbonate transfer agents used in RAFT [85]

ed its use by making a pSt-pnBA-pSt ABA triblock copolymer with M_n=161,500 and M_w/M_n=1.16. The GPC traces indicated clean chain extension with little low molecular weight tailing that would indicate unreacted macroinitiator [85]. One aspect of the RAFT systems that differs from nitroxide and ATRP systems is that block copolymers prepared using this RAFT agent will grow from the inside out, rather than by chain extension of the center block, as is typical for other CRP methods. Block order is also important in these systems. Since the transfer constant for methacrylate monomers is lower than for either the acrylate or styrene monomers, inefficient blocking may occur if the methacrylate monomer is the second block polymerized [53]. This is similar to what occurs in the ATRP systems; however, the option of halogen exchange is not available to overcome it.

First block copolymers by degenerative transfer technique were prepared by using alkyl iodides/AIBN in polymerization of BA and St. For example, at 70 °C pSt with M_n=2500 and M_w/M_n=1.45 was extended to pSt-b-pBA with M_n=8630 and M_w/M_n=2.20. Although polydispersities were relatively high, a clean shift of the entire MW distribution was observed. In a similar way pBA with M_n=2820 and M_w/M_n=1.60 was extended to pBA-b-pSt with M_n=6290 and M_w/M_n=1.75. Conversion of the first block was above 95% and the second ~80% in both cases [51].

RAFT can also be used to directly incorporate acrylic or methacrylic acid into block copolymers [53], i.e., without the need for protecting groups, something that cannot yet be accomplished using ATRP or nitroxide systems. TMPAH was used to prepare copolymers of acrylic acid, but no block copolymers were reported [71]. Successful polymerization of the salts of some acidic monomers has been accomplished using ATRP in aqueous media, which, as detailed above, eliminated the need for an additional deprotection step [214]. Block copolymers incorporating DMAEMA were also prepared using RAFT, by chain extension of a poly (benzyl methacrylate) macroinitiator (M_n, block= 3500, M_w/M_n=1.06) [85]. However, very long reaction times were necessary due

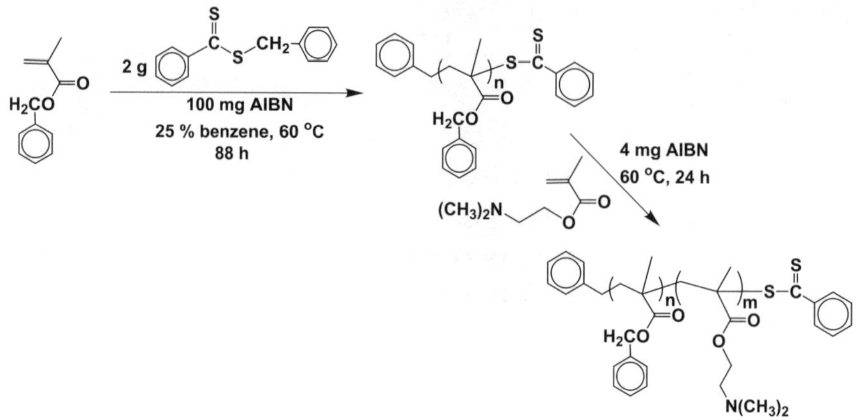

Scheme 20. Synthesis of p(BzMA-b-DMAEMA) block copolymers using RAFT [85]

to the significant retardation at high concentrations of the RAFT reagent, as shown in Scheme 20.

A pSt macroinitiator of $M_n=20,300$ with $M_w/M_n=1.15$ was chain extended with DMA, resulting in a block copolymer with $M_n=43,000$ and an $M_w/M_n=1.24$ [85]. The authors did not present monomer conversion data or GPC traces, so no conclusion about the rate of the reaction or blocking efficiency can be drawn; however, there was no indication in the text that the polymerization was uncontrolled, as was originally observed for the ATRP system [200].

Block copolymers of St and BA were also prepared using miniemulsion polymerization with fluoroalkyl iodides as transfer agents. The best results were obtained using a slow monomer feeding technique to reduce the propagation rate and enhance the rate of exchange reactions [212, 215].

3.1.4
Comparison of CRP Methods for Block Copolymer Synthesis

The above sections provide information about the synthesis of block copolymers using CRP techniques. In the first section, details about the nitroxide-based systems were presented. From the literature reports, it is clear that TEMPO and TEMPO-based analogs only produce block copolymers of St-based monomers successfully. Liquid crystalline monomers, sugar containing derivatives, and even silicon-based St monomers were successfully homopolymerized. Water soluble monomers like sodium 4-styrenesulfonate and 4-(dimethylamino) methylstyrene were successfully incorporated into block copolymers using TEMPO in aqueous media. However, there was evidence that even with some St derivatives, particularly those that were chloromethylated, side reactions occurred to produce dead polymer chains when TEMPO was used to moderate the polymerization. Attempts at using the system for *n*-butyl acrylate, isoprene, butadiene, or vinyl pyridine produced polymers with low chain end functionality and poorly defined block copolymers. This can be attributed to differences in the equilibrium conditions for these monomers as opposed to St. New nitroxides have overcome this problem and have allowed for the homo- and copolymerizations of various acrylate, acrylamide, and diene monomers. While most investigations have focused on the statistical copolymerizations, block copolymers of St with isoprene and *tert*-butyl acrylate as well as isoprene with *tert*-butyl acrylate have been synthesized. Unfortunately, chain end functionality is lost if the acrylate monomer is the second block in either chain extension, potentially limiting application of the system. No controlled homopolymers or block copolymers containing methacrylates have yet been prepared. Further alteration in the nitroxide structure may eventually allow for the incorporation of methacrylate monomers; however, this presently remains elusive in the nitroxide realm.

ATRP can be carried out in either organic or aqueous media, as well as under biphasic conditions. Monomers that can be incorporated into block copolymers with ease include styrenes, acrylates, methacrylates, and even vinyl pyridine. Some problems associated with preparation of block copolymers from less reactive macroinitiators and methacrylate-type monomers can be overcome by using the halogen exchange technique. This allows for a better match between the rates of cross-propagation and polymerization of the second monomer, enhancing the physical properties of the block copolymers. Polymers can be prepared using an isolation/purification procedure for the macroinitiator as well as a sequential monomer addition technique. The first approach produces "clean" block copolymers while the latter approach creates a gradient of monomers in the second block. MMA/nBA copolymers produced using both techniques had different mechanical properties, providing another route to altering materials for specific applications.

There are numerous combinations of transition metal and ligand that can be used to tailor the ATRP catalyst system to specific monomers. The ATRP systems are tolerant of many impurities and can be carried out in the presence of limited amount of oxygen and inhibitors [216, 217]. This approach is so simple that it has been proposed as undergraduate experiments to prepare block copolymers [218, 219]. However, the ATRP catalyst can be poisoned by acids, but the salts of methacrylic and vinylbenzoic acids have been polymerized directly in aqueous media [206]. Also, the use of protecting groups [206], followed by a deprotection step to yield the acids, has been successful in organic media [220]. While ATRP cannot be used to prepared well-defined polymers of vinyl acetate [221] as of yet, these goals may be realized with further catalyst development.

Block copolymers prepared using the degenerative transfer with alkyl iodides and RAFT systems include methacrylate, acrylate, and styrene combinations as well as styrene/acrylate, acrylate/acrylic acid, methacrylate/styrene, and methacrylate/methacrylic acid combinations. A pSt macroinitiator was also successfully chain extended with *N,N*-dimethylacrylamide in a controlled fashion. While this system is useful for a variety of combinations of monomers, the order in which the individual blocks are produced is important. Those monomers with higher transfer rates must be polymerized first to avoid inefficient blocking when chain extended with monomers with lower transfer rates. Although an approach similar to that of the halogen exchange would solve this problem, this has not yet been addressed for the RAFT systems and may limit its applicability for materials synthesis.

3.2
Block Copolymers Prepared Through Transformation Techniques

There have been numerous examples of transformation reactions that combine either preformed macroinitiators or different polymerization techniques with the CRP of various monomers reported in the literature.

The first section focuses on using commercially available polymers that were transformed into controlled radical macroinitiators. Commercial availability and the use of simple organic chemistry to prepare the macroinitiator make this route to block copolymers attractive to many. The second section focuses on those examples where two different polymerization methods were combined. After obtaining a functional molecule through one process transformation to the species suitable for controlled radical polymerization may or may not be necessary, as detailed below. Although historically the first example of a transformation without altering the chain end but just changing the catalytic system was reported by Coca and Matyjaszewski and involved the cationic polymerization of styrene followed by the ATRP of styrene [222], and the first incorporation of bromoesters to a hydroxy-terminated polymer was reported for polysulfones followed by the ATRP of St and nBA [223], the following sections are presented in a way that focuses on the methods used rather than the historical perspective.

3.2.1
CRP from Commercially Available Macroinitiators

3.2.1.1
Poly(ethylene glycol)

The majority of the work reporting on use of commercially available macroinitiators to prepare block copolymers describes the use of macroinitiators derived from poly(ethylene glycol) (PEG), which can be used to produce amphiphilic block copolymers. These polymers contain hydroxy end-functionalities that can

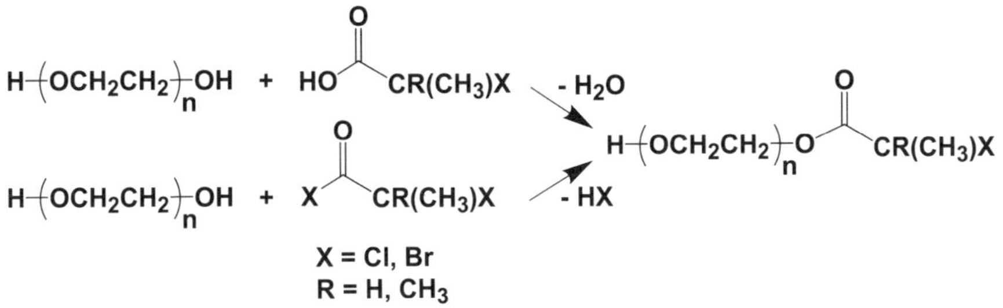

3 Linear Block Copolymers

$$H{-}(OCH_2CH_2)_n{-}OH + HO{-}CO{-}CR(CH_3)X \xrightarrow{-H_2O}$$

$$H{-}(OCH_2CH_2)_n{-}OH + X{-}CO{-}CR(CH_3)X \xrightarrow{-HX} H{-}(OCH_2CH_2)_n{-}O{-}CO{-}CR(CH_3)X$$

X = Cl, Br
R = H, CH$_3$

Scheme 21. General procedures for modifying hydroxy-functional PEG to make ATRP macroinitiators

easily be altered to incorporate CRP-type initiating end groups through simple chemistry, as illustrated in Scheme 21.

Kops et al. first demonstrated the use of PEG macroinitiators [224]. They prepared a 2-bromo- or 2-chloropropionate difunctional PEG macroinitiator (M_n= 2180 or 2270, respectively) and initiated the polymerization of St with the corresponding CuX/bpy complex (where X=chlorine (130 °C) or bromine (110 °C)) [224]. The block copolymers prepared using the chlorine system had M_n=19,700 and M_w/M_n=1.29 (bulk) or M_n=4700 with M_w/M_n=1.28 (80 vol.% xylene solution). There was little evidence in the block copolymers of the presence of a homopolymer of St resulting from a thermal polymerization. Attempts to prepare the same block copolymer using PEG-TEMPO as the macroinitiator resulted in contamination with pSt due to the similar rates for the polymerization of styrene whether initiated by the macroinitiator or by the thermal process [225]. The pSt content in the block copolymers using the ATRP process was slightly higher than that predicted by the monomer conversion, as determined by ^{1}H NMR analysis. The GPC traces showed clean chain extension from the macroinitiator for the polymers prepared both in bulk and in solution, indicating the functionalization of the PEG was complete and the blocking efficiency was high [224]. Detailed analysis performed later showed that initiation wasn't complete until >50% monomer conversion but that no remaining macroinitiator was found when the block polymer was extracted with water [226]. Cheng et al. later reported on an identical chlorine-based system, but did not investigate the rate of consumption of the macroinitiator [227].

Kubisa et al. also used hydroxy-functional PEG after reaction with 2-bromopropionyl bromide as an ATRP macroinitiator [228]. Their goal, however, was to polymerize *tert*-butyl acrylate, rather than St, then to hydrolyze the esters to acid functionality and study the cation binding properties of the doubly amphiphilic block copolymers. They utilized a CuBr/PMDETA catalyst system for the polymerization and, although the macroinitiator was completely consumed, MALDI-TOF analysis indicated that bromine was replaced with a hydrogen at

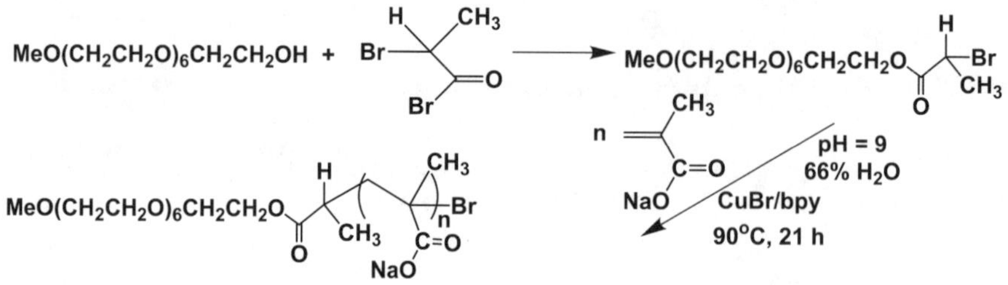

Scheme 22. Transformation of a PEG oligomer to an ATRP initiator followed by chain extension via ATRP with the sodium salt of methacrylic acid [214]

monomer conversions >75% [229]. However, these experiments were carried out in the presence of an excess of the PMDETA ligand, which can act as a transfer agent. Davis and Matyjaszewski showed that choosing the proper ATRP conditions (i.e., not using an excess of PMDETA) resulted in good chain end functionality and high blocking efficiency [196].

Armes et al. used a pEG macroinitiator to polymerize the sodium salt of methacrylic acid to form a block copolymer with a short PEG segment, as shown in Scheme 22 [214]. Critical for success was pH control. Too low a pH caused protonation of the ligand and too high a pH caused nucleophilic displacement of Br from the chain ends.

Ying et al. used a dihydroxy functional poly(propylene oxide) of M_n=2000 (M_w/M_n=1.28) as a macroinitiator for the ATRP of St after esterification with chloroacetyl chloride [230]. The polymerization using the CuCl/bpy catalyst system had an induction period (~3 h), then proceeded with a constant concentration of active species. During this induction period, the GPC traces indicated slow consumption of the macroinitiator, corresponding to the higher than expected molecular weights and a significant increase in the M_w/M_n to 2.5. After >20% monomer conversion, the molecular weights increased linearly ($M_{n,final}$= 11,900) and M_w/M_n decreased with increasing conversion ($M_w/M_{n,final}$=1.58) [230]. These results were similar to those found by Kops et al. [224].

Moad et al. functionalized a monomethylated PEG with a dithioester to prepare a RAFT macroinitiator (M_n=750, M_w/M_n=1.04). The formation of a block copolymer with either St (M_n=7800, M_w/M_n=1.07) or with benzyl methacrylate (M_n=10,800, M_w/M_n=1.10) resulted in well-defined copolymers with no remaining PEG macroinitiator [53].

3.2.1.2
Biocompatible Polymers

Yoshida and Nakamura reacted poly(ethylene adipate) (EAD) containing isocyanate end groups with HTEMPO to produce a counter radical (Scheme 23) al-

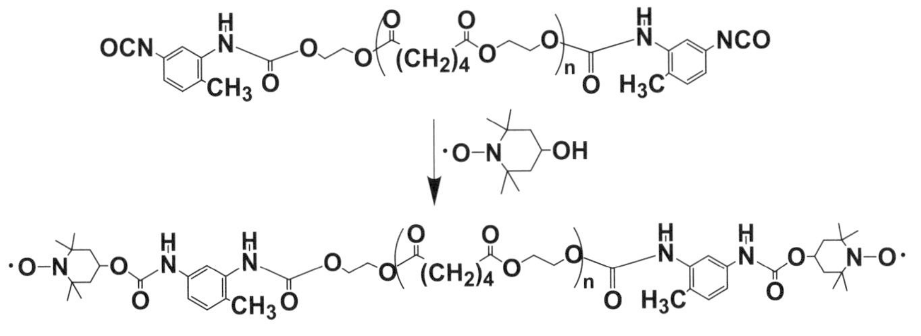

Scheme 23. Synthesis of TEMPO-terminated poly(ethylene adipate) [231]

lowing for a copolymerization with St in the presence of BPO to form a block copolymer [231].

The presence of the nitroxide radical was confirmed through EPR and ^{1}H NMR spectroscopic methods. The copolymer GPC trace (M_n=33,100, M_w/M_n=1.37) was symmetrical with no evidence of unreacted macroinitiator or homopolymer of St resulting from either thermal initiation or from disproportionation of the pSt from the TEMPO chain end [231]. The kinetic results showed a first-order relationship between monomer conversion and time and the molecular weights increased linearly with conversion, indicating the polymerization proceeded with minimal termination or chain transfer reactions. The presence of the pEAD block produces an amphiphilic copolymer with a biodegradable block that may be useful for biomedical applications.

Haddleton and Ohno acylated and ring-opened β-cyclodextrin (β-CD), which was subsequently selectively deacylated to create a hydroxy functional maltoheptose [204]. After reaction with 2-bromoisobutyryl bromide, the new initiator was used for ATRP of MMA using the CuBr/n-propyl-2-pyridylmethanimine catalyst system. The polymerization occurred with little termination and a linear increase of the molecular weight with time (final M_w/M_n=1.09) [204]. After complete polymerization, the maltoheptose unit was deprotected to form the amphiphilic block copolymer. Several other methacrylate derivatives were also used for the preparation of block copolymers, including DMAEMA, a poly (ethylene glycol) macromonomer (pEGMA), and a sugar-containing monomer. Although the block copolymer with DMAEMA could not be characterized by GPC, the other copolymers had predictable molecular weights and low polydispersities (M_w/M_n<1.25) [204]. These are interesting copolymers because they provide a route to functionalized oligosaccharides.

Similarly, poly(dimethylsiloxane) (pDMS), an inorganic polymer, has been incorporated into block copolymers using a functionalization reaction. Matyjaszewski et al. used commercially available hydrosilyl or vinylsilyl terminated

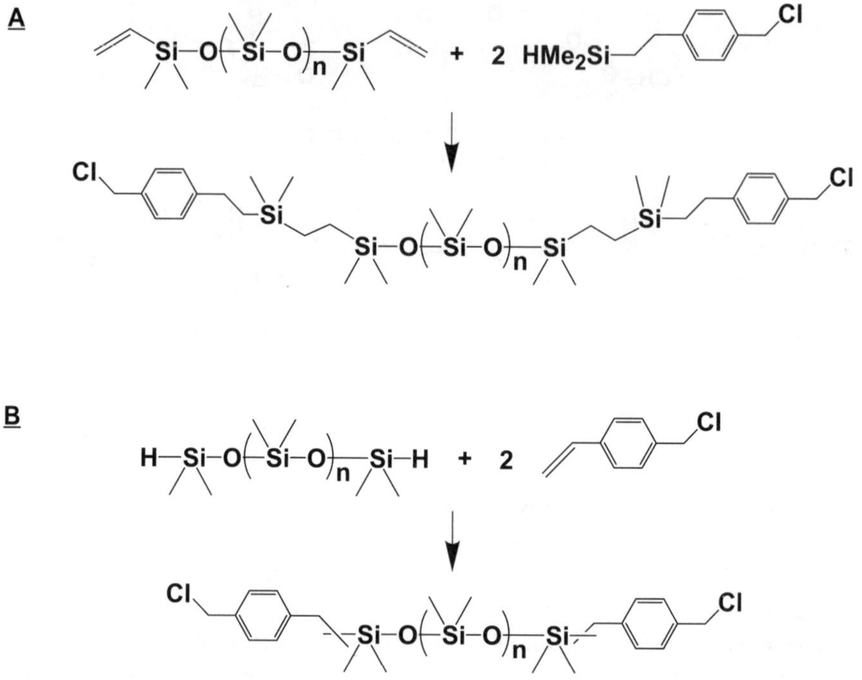

Scheme 24. A Synthesis of benzyl chlorine-based ATRP macroinitiators from: A vinylsilyl-; B hydrosilyl-terminated PDMS [232, 233]

pDMS functionalized with a benzyl chloride moiety to produce macroinitiators for ATRP reactions (Scheme 24) [232, 233].

The resulting ABA block copolymer using the macroinitiator originating from the vinylsilyl pDMS with St was more well-defined than the commercial pDMS sample (M_n=9800, M_w/M_n=2.40), with an M_n=20,700 and an M_w/M_n=1.60. Better results were obtained from the hydrosilyl derived macroinitiator, with the molecular weight increasing from M_n=3000 to 9900 and M_w/M_n decreasing from 1.35 to 1.19 [233]. Subsequently, pDMS was prepared anionically, functionalized, and used as an ATRP macroinitiator. This will be detailed in the following section on transformation reactions [234].

3.2.1.3
Polydienes

Nitroxide-mediated polymerizations have recently been used to polymerize diene-based monomers [159]. ATRP [79] has so far been less successful and therefore other routes to incorporation have been utilized. One route is to functionalize a commercially available olefin based polymer. Kops et al. reacted a hy-

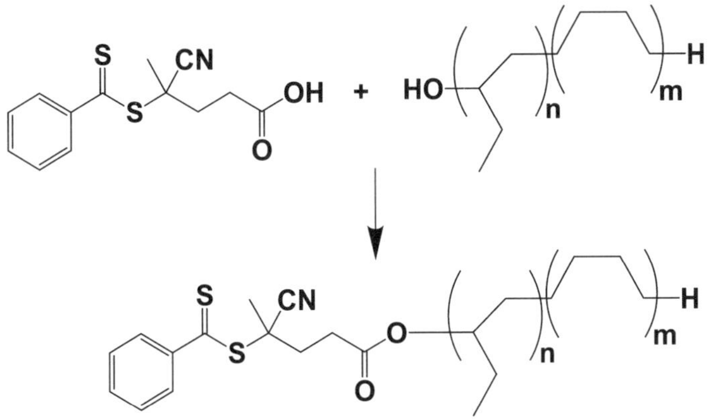

Scheme 25. Reaction of p(EB) with 4-cyano-4-[(thiobenzoyl)sulfanyl] pentanoic acid to produce a RAFT agent [134]

droxy-functional poly(ethylene-co-butylene) (EB) with bromopropionyl chloride to produce a bromo-terminated ATRP macroinitiator [235]. Using the Cu-Br/bpy catalyst system, the resulting block copolymers with St had molecular weights ranging from M_n=18,600 to 24,100 with M_w/M_n=1.29 to 1.36. The GPC traces were symmetrical with little evidence of unreacted macroinitiator, even though the commercial sample had a hydroxy functionality of 0.98. DSC analysis showed the presence of two T_gs, one at −63 °C (pEB) and the other at 93 °C (pSt), supporting the idea that chain extension had occurred [235].

The same hydroxy-functional pEB was reacted with 4-cyano-4-[(thiobenzoyl)sulfanyl] pentanoic acid to produce a macroinitiator for a RAFT copolymerization (Scheme 25) [134]. Klumperman et al. used this RAFT agent to initiate the polymerization of St or the copolymerization of St with MAh.

The molecular weight of the pSt-b-pEB copolymer increased linearly with monomer conversion with the final M_n=23,000 and the M_w/M_n=1.20. There was, however, a large tail to lower molecular weights comprised of a small proportion of pSt derived from the azo initiator used to start the transfer process as well as what appears to be unreacted macroinitiator, although the authors did not discuss this fully [134]. When low molecular weight block copolymers were subjected to UV irradiation, the end groups were cleaved and the color associated with the thioester end groups was easily removed by filtering through a silica column. The GPC traces showed evidence of coupling products associated with ABA block copolymer formation as well as three-arm stars, the result of the polymer chain end radical coupling to a RAFT intermediate. The same pEB macroinitiator was used for the RAFT-based copolymerization of St with MAh; however, the results of this polymerization indicated that the growth from the macroinitiator was not uniform because of slow initiation, regardless of the low molecular weight distribution (M_w/M_n=1.12) [134]. Similar work was carried out by ATRP, confirming good functionalities and block extension [236].

Ying et al. transformed a dihydroxy-terminal pBD sample (M_n=6540, M_w/M_n=2.1 to 2.3) into an ATRP macroinitiator through an esterification reaction with chloroacetyl chloride [230]. The block copolymer had an M_n=19,000 with an M_w/M_n=2.16. Although specific kinetic results were not shown for this system, an induction period similar to the one seen for the pPO ATRP macroinitiator [230] would be expected.

Table 7. Commercially available macroinitiators transformed into CRP initiators.

Polymer Structure	Name	CRP Method
	poly(ethylene glycol)	ATRP[a], RAFT[b]
	poly(propylene oxide)	ATRP[c]
	poly(ethylene adipate)	TEMPO[d]
	poly(β-cyclodextran)	ATRP[e]
	poly(dimethylsiloxane)	ATRP[f]
	poly(ethylene-co-butylene)	ATRP[g], RAFT[h]
	poly(butadiene)	ATRP[i]

[a]Kops[224, 226], [b]Moad[53], [c]Ying[230], [d]Yoshida[231], [e]Haddleton[204], [f]Matyjaszewski [232, 233], Haddleton [371], [g]Kops [235], Klumpermann [236], [h]Klumpermann [134], [i]Ying [230]

3 Linear Block Copolymers

3.2.1.4
Summary

The above examples describe the transformation of commercially available polymers into CRP macroinitiators to prepare unique materials and illustrate just some of the possibilities that are open to exploration. The ease with which hydroxy functionalities in particular can be incorporated into a polymer and then transformed into CRP initiating moieties provides a route to modify many already existing materials rather than having to develop completely new materials for a specific application. For example, PEG derivatives are often used as surfactants; however, the hydrophobic tails are generally long alkyl chains. Chain extending PEG with a particular monomer via CRP may result in surfactants that are either more versatile or more specific for a certain application. The same can be said for the incorporation of biocompatible segments into polymers, particularly for drug delivery applications. Table 7 contains a summary of the above discussion on these polymers.

3.2.2
Block Copolymers by Combination of Two Polymerization Techniques

Many polymerization techniques have been combined with CRP through site transformation of the active species. These include non-living techniques like condensation (or step) and conventional free radical processes or living methods like anionic, cationic, and ring-opening polymerizations, as well as others. Early examples were undertaken perhaps just to show that two different techniques could be combined, while later examples show how elegant the combinations have become and provide a foundation for controlled synthesis of materials from any type of monomers. These types of reactions are detailed below.

3.2.2.1
CRP and Non-Living Polymerizations

Non-living polymerization techniques can be combined with CRP methods to produce block copolymers. The first example of transforming a hydroxy functionality into an ATRP initiator was demonstrated by Gaynor and Matyjaszewski [223], who converted a polysulfone, prepared through the condensation polymerization of 4,4-difluorosulfone with an excess of bisphenol A, to an ATRP initiator by reaction with 2-bromopropionyl bromide for subsequent controlled polymerization reactions (cf. Scheme 26). The transformation proved to be quantitative and the macroinitiator (M_n=4030, M_w/M_n=1.5) was used for formation of triblock copolymers with St (M_n=10,700, M_w/M_n=1.1) or nBA (M_n=15,300, M_w/M_n=1.2) as shown in Fig. 31 [223]. DSC analysis provided evidence of the presence of two distinct blocks with T_g=153–159 °C (polysulfone) and 104 °C (pSt) or –41 °C (pBA).

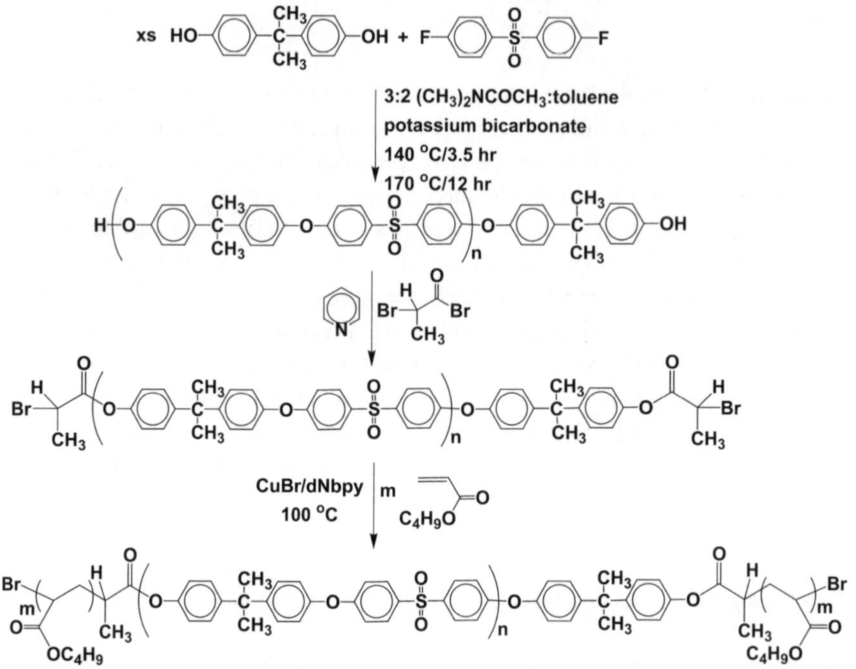

Scheme 26. Synthesis of polysulfone-b-pnBA by transformation of a macroinitiator prepared through a condensation polymerization to an initiator for ATRP [223]

The triblock copolymer with a central polysulfone segment (25 wt%) and pBA side blocks organizes into supramolecular aggregates with a periodicity from 10 to 12 nm. According to SAXS, the periodicity remains even above 250 °C, although DMA indicates that the triblock copolymer "melts" at about 100 °C. This temperature corresponds to a structural relaxation of pnBA with a molecular weight of a few million, confirming a high degree of aggregation (Fig. 32).

Similar to the synthesis of the difunctional polysulfone macroinitiator, a polyester was used in the synthesis of block copolymers by ATRP. The α, ω-dihydroxy terminal polymer was synthesized by the transesterification of 1,6-hexanediol with dimethyl adipate [237]. The end groups were then esterified with 2-bromopropionyl bromide and the ATRP of styrene yielded the ABA triblock copolymers.

Recently, the synthesis of rigid-flexible triblock copolymers, with a rigid central part and possessing photoluminescence, has been described [238, 239]. First, Suzuki coupling was applied to prepare α,ω-acetoxy functionalized oligophenylenes with five or seven rings. Hydrolysis of these acetoxy end groups and esterification of the resulting hydroxy end groups with acyl chlorides led to molecules capable of acting as ATRP initiators. The final rigid-flexible copolymers of styrene displayed low polydispersities and showed blue light emission.

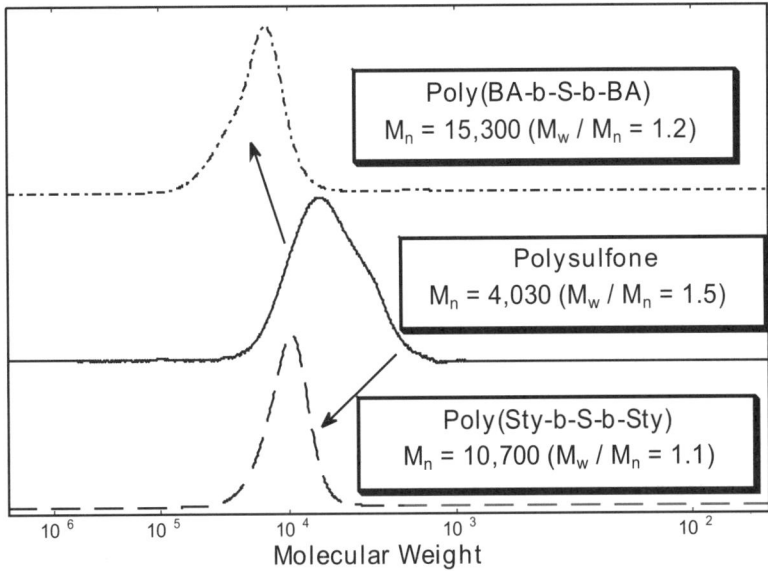

Fig. 31. GPC traces for polysulfone, pSt-b-polysulfone-b-pSt, and pnBA-b-polysulfone-b-pn-BA. Reprinted with permission from [223]. Copyright (1997) American Chemical Society.

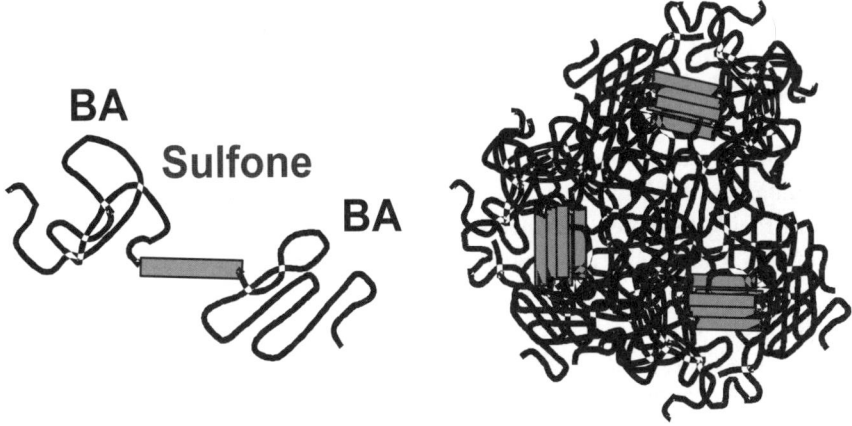

Fig. 32. Illustration of the aggregation between phase separated blocks of polysulfone and pnBA

Jones et al. prepared poly(methylphenylsilylene) (p(MPSi)) using a Wurtz-type reductive coupling reaction, followed by reaction with (4-chloromethylphenylethyl) dimethylchlorosilylene to produce the α,ω-difunctional ATRP initiator [240]. ^{1}H NMR characterization indicated the presence of several different types of chains ends; however, the majority of them appeared to contain the desired chlorine. Chain extension of the macroinitiator (M_n=4550, M_w/M_n=5.9, M_p=3100) with St using the CuBr/bpy catalyst system resulted in a block copolymer with a bimodal distribution. The lower molecular weight peak corresponded to dead macroinitiator (11 wt%) while the higher molecular weight peak was a combination of both AB (M_n=206,600, M_w/M_n=1.91, 50 wt%) and ABA (M_n=527,100, M_w/M_n=1.43, 39 wt%) block copolymers [240]. While the block copolymers were more well-defined than the macroinitiator, the mixture of mono- and difunctional chains may affect the properties of the material and limit its applicability in the electronics industry.

Kallitsis et al. used a Suzuki coupling reaction to prepare α, ω-acetoxy functionalized oligophenylenes, which were then converted to the ATRP macroinitiators through hydrolysis, followed by esterification with acyl chlorides [241]. The macroinitiators had 4 (quatra phenyl, QP) or 6 (heptaphenyl, HP) phenyl units in the backbone which emanated from an aromatic core initiator. The QP-pSt block copolymers had M_n=7900 to 20,600 with M_w/M_n=1.17 to 2.15. The low polydispersity polymers were prepared using the CuBr/bpy catalyst system, while the samples with larger M_w/M_n were the product of either a $CuBr_2$/Cu(0)/bpy or a CuBr/Cu(0)/bpy catalyst system. The rate enhancement offered by the addition of Cu(0) [242] resulted in the generation of too many radicals and an ill-defined product. The HP-pSt block copolymers had M_n=2100 to 8000 with M_w/M_n from 1.19 to 1.41. Use of a heterogeneous catalyst system resulted in a sample with M_w/M_n=1.41; all the other polymerizations used the CuBr/bpy system [241]. The QP-pSt polymers exhibited a red shift from 378 nm to 405 nm, suggesting that the presence of the pSt block may be a way to tailor the properties of this type of rod-coil polymers.

To enhance the photovoltaic efficiency of poly (phenylenevinylene) (PV)-C_{60}, Hadziioannou et al. synthesized block copolymers by combining condensation and nitroxide-mediated polymerizations [243]. Oligo pPV with an ω-aldehyde group was reacted with the Grignard reagent containing the TEMPO moiety capable of initiating the polymerization of St. The average degree of polymerization of the pPV was 7 (M_n=2500) and chain extension produced a block copolymer of M_n=7000 with a weight ratio of the pSt block to the pPV block of 1.8. In order to incorporate the C_{60} group, a statistical copolymerization of St with CMSt was performed at a feed ratio of 2:1 St:CMS to generate a block copolymer with M_n=9000 [243]. Subsequently, the block copolymer was functionalized with C_{60} through an ATRA reaction with the chloromethyl groups in the backbone (Scheme 27). Larger amounts of CMSt in the feed led to crosslinked products once functionalization was carried out. The C_{60} containing block copolymers had, on average, 15 C_{60} units incorporated. Electron transfer studies showed that the photoluminescence of the pPV segment is almost entirely quenched once the C_{60} is added, suggesting that this material may find utility for charge transfer media [243].

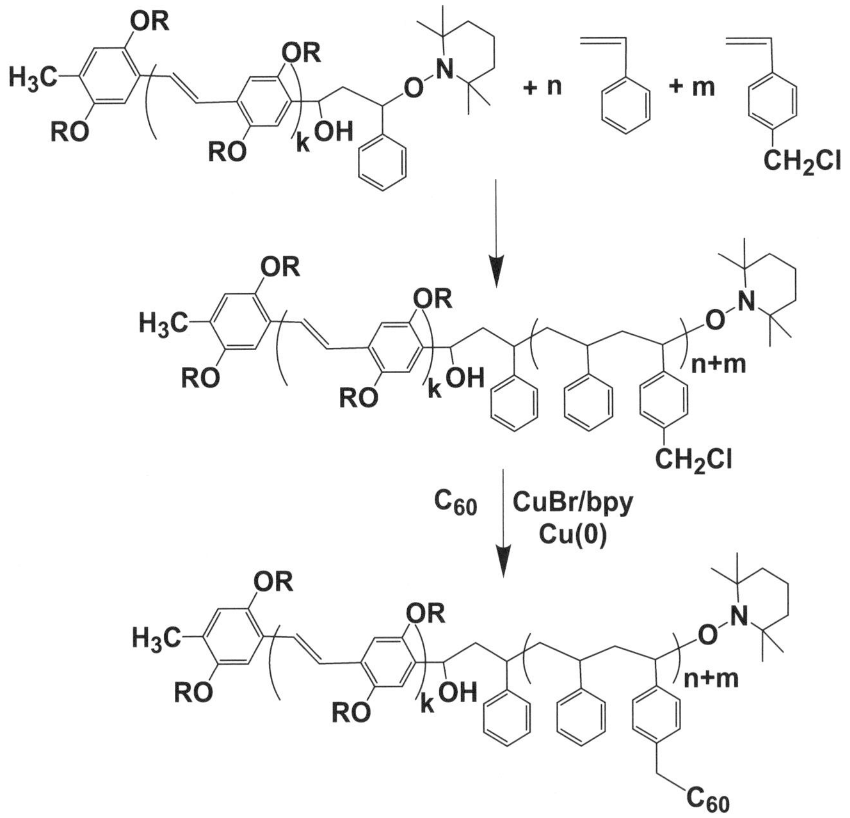

Scheme 27. Preparation of a block copolymer of pPV with St and CMSt using TEMPO-mediated polymerization, followed by ATRA to incorporate C_{60} [243]

Yoshida and Tanimoto used pDMS that contained, on average, five azo moieties in the main chain as a macroinitiator for St in the presence of MOTEMPO [245]. The initiator efficiency was 39%, based on trapping studies performed with MOTEMPO and followed by UV detection. The macroinitiator was used for the polymerization of St, which was carried out in bulk at 130 °C, with 84% monomer conversion after 72 h (M_n=66,000, M_w/M_n=1.90). There was no unreacted pDMS macroinitiator present in the final product, suggesting that any azo groups not participating in the chain extension were deactivated through a coupling reaction. In addition, only when the ratio of MOTEMPO to the azo initiator was 0.6 were monomodal GPC traces obtained; with increasing amounts of MOTEMPO, the contribution of the thermal polymerization began to become significant [245].

Further chain extension of the pDMS-b-pSt with p-methoxystyrene produced a block copolymer with increased molecular weight (M_n=135,000) and a still narrow molecular weight distribution (M_w/M_n=1.43) [245]. Some tailing in the

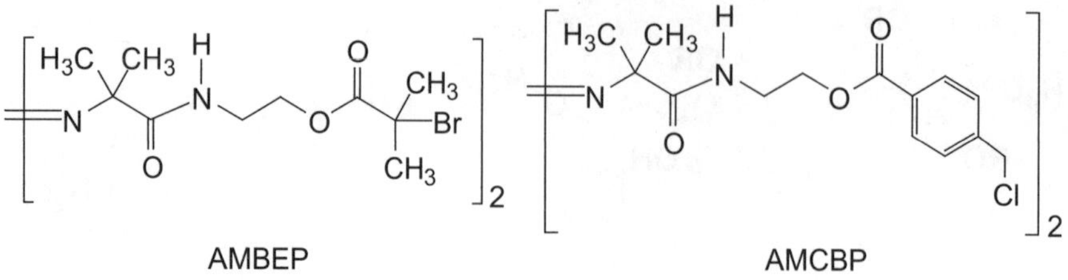

Fig. 33. Structure of 2,2′-azobis[2-methyl-N-(2-(2-bromoisobutyryloxy)ethyl)-propionamide (AMBEP) and 2,2′-azobis[2-methyl-N-(2-(4-chloromethylbenzoyloxy)-ethyl)-propionamide (AMCBP) [246]

GPC trace was evident, which would indicate that the blocking efficiency may not have been 100%; however, these experiments show that the use of a macroazo initiator can be successfully combined with CRP techniques.

Using a similar methodology, Matyjaszewski et al. modified a diazo initiator containing terminal hydroxy groups with either a 2-bromoisobutyryloxy group (AMBEP) or a 4-chloromethylbenzoyloxy group (AMCBP) to produce a compound capable of initiating both a free radical polymerization and an ATRP reaction (Fig. 33) [246]. Their goal was to incorporate vinyl acetate (VA), which cannot yet be accomplished solely using ATRP, into a block copolymer. Four different combinations of reactions were examined.

The first route utilized AMBEP to polymerize VA; however, transfer reactions with the bromine moiety caused low monomer conversion and formed low molecular weight polymer. When the AMCBP was used as the initiator, transfer was limited and a polymer with M_n=47,900 and M_w/M_n=2.21 was formed. Formation of a block copolymer by polymerization of St, via the ATRP initiating sites, resulted in polymers with increased molecular weights (M_n=91,600) and narrower molecular weight distributions (M_w/M_n=1.80) [246]. The second route to VA incorporation was to use the alkyl halide initiator for the ATRP of acrylates at room temperature first, followed by the conventional radical polymerization of VA. The pnBA macroinitiator prepared using the CuBr/tris [2-dimethylaminoethyl] amine catalyst system had an M_n=7500 with an M_w/M_n=1.15. (cf. Fig. 34). Unfortunately, the blocking efficiency using this macroazo initiator was only about 50%, resulting in the formation of a polymer showing contamination from both in cage coupling and disproportionation reactions [246]. Finally, two methods based on using redox initiated polymerizations in conjunction with ATRP were attempted. In the first approach a pVA with M_n=3600 and an M_w/M_n=1.81 was prepared using CCl_4 as an initiator/transfer agent in the presence of $Fe(OAc)_2$/PMDETA as a catalyst and was chain extended with St to yield a block copolymer with M_n=24,300 and M_w/M_n=1.42 or with nBA to produce a block copolymer with M_n=11,000 and M_w/M_n=1.41 as shown in Fig. 34. Alterna-

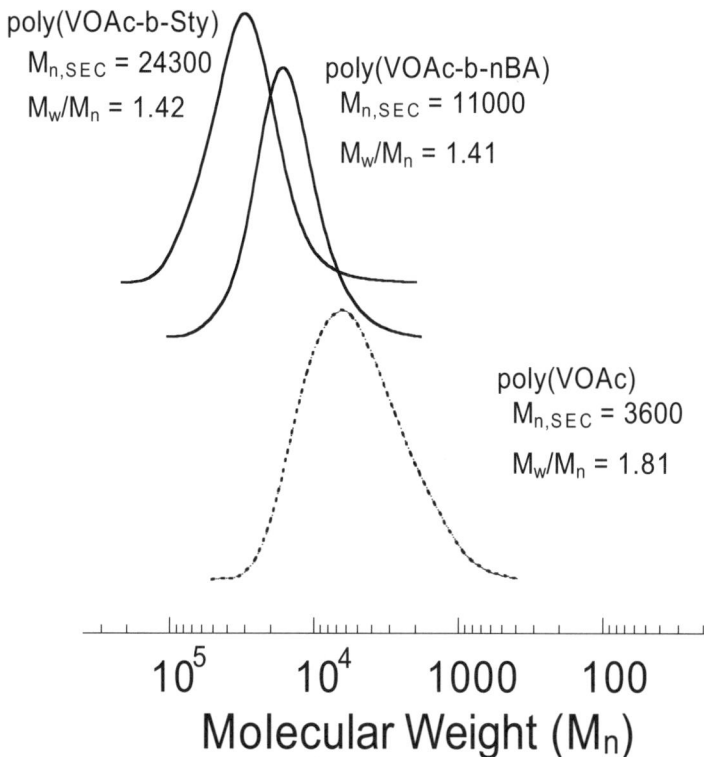

Fig. 34. GPC chromatograms of pVA prepared using $CCl_4/Fe(OAc)_2/PMDETA$ and its subsequent block copolymerizations. Conditions for the synthesis of pVA: 50°C; $[VA]_0=5.4$ mol/l (50vol% ethyl acetate); $[VA]_0/[CCl_4]_0=40$; $[CCl_4]_0/[Fe(OAc)_2]_0/[PMDETA]_0=1/0.2/0.2$; time=10h; conversion=45%; $M_{n,SEC}=3600$, $M_w/M_n=1.81$. Conditions for the synthesis of p(VA-b-pSt): 90°C; $[St]_0=8.7$ mol/l (bulk); $[St]_0/[pVA-CCl_3]_0=628$; $[p(VA)-CCl_3]_0/[Cu-Br]_0/[dNbpy]_0=1/2.5/5.0$; time=15h, conversion=46%. Conditions for the synthesis of p(VA-b-nBA): 90°C; $[nBA]_0=7.0$ mol/l (bulk); $[nBA]_0/[pVA-CCl_3]_0=502$; $[pVA-CCl_3]_0/[Cu-Br]_0/[dNbpy]_0=1/2.5/5.0$; time=11h, conversion=15%. Reprinted with permission from [246]. Copyright (1999) American Chemical Society.

tively, the ATRP of nBA was first carried out ($M_n=2460$, $M_w/M_n=1.33$), followed by the redox polymerization of VA ($M_n=4450$, $M_w/M_n=2.58$) [246]. All four methods are useful since they require no transformation chemistry to obtain the macroinitiators.

Boutevin et al. also incorporated VA into block copolymers, but via a slightly different technique. They utilized chloroform as a transfer agent, which, in the presence of AIBN as an initiator, will produced telomers of VA that are capable of initiating ATRP [247]. Various telomers of pVA (degree of polymerization=1, 9, or 62) were used as the macroinitiators for the ATRP of St, carried out in the presence of the CuCl/bpy catalyst system. The molecular weights increased linearly with increasing monomer conversion (average final M_n~8000) and the molecular weights decreased throughout the polymerization with final M_w/M_n<1.5. The content of pVA in the polymer calculated from ^{1}H NMR analysis agreed well with the theoretical composition [247]. In addition, analogous to the procedure used by Matyjaszewski et al. [246], Destarac and Boutevin modified a diazoinitiator with trichloroethanol to produce an initiator capable of both a free radical polymerization and ATRP [248]. After polymerizing nBA free radically, the macroinitiator was used for the ATRP of St with the CuCl/bpy catalyst system. The ATRP reaction had a constant concentration of active species during the polymerization and a linear increase of the molecular weight with conversion ($M_{n,final}$=22,620). The molecular weight distribution decreased from M_w/M_n= 2.32 to M_w/M_n=1.66 by the end of the copolymerization. It was assumed that the predominant mode of termination in the macroinitiator was coupling and that the block copolymers produced were of the ABA type [248].

Using a conventional radical process, Ying et al. prepared difunctional p(vinylidene fluoride) (VDF) in the presence of 1,2-dibromotetrafluoroethane as a chain transfer agent [249]. The ATRP macroinitiator had an M_n=3870 with an M_w/M_n=1.1 and a bromine-functionality close to 2. Chain extension of the pVDF macroinitiator with St using the CuBr/bpy catalyst produced a polymer with M_n=37,360 and with M_w/M_n=1.65. The experimental molecular weights, although they increased linearly with conversion, were significantly higher than the theoretical values, which may have resulted from poor solubility of the pVDF in the GPC mobile phase. The GPC traces indicated complete consumption of the macroinitiator and ^{1}H NMR analysis gave a composition of 14% VDF and 86% St in the copolymer [249].

Destarac and Matyjaszewski et al. showed that α-trichloromethylated pVDF telomers could be used as ATRP macroinitiators to prepare block copolymers with various monomers [250]. The polymerization of VDF was carried out using a peroxide initiator in the presence of chloroform, which produced an α-trichloromethyl radical capable of adding to the monomer. The subsequent ATRP of St, MMA, MA, or nBA using these macroinitiators resulted in block copolymers with M_n>6000 and M_w/M_n<1.3. GPC traces verified complete consumption of the macroinitiators [250].

In a novel twist, Yagci et al. used an ATRP initiator capped with a thiophene moiety to polymerize MMA. The resulting macromonomer was then used for the electropolymerization of pyrrole (PY) (Scheme 28) [251]. N-Trichloroacetyl 3-thiophenyl methyl carbamate was prepared through the reaction of 3-thiophenemethanol with trichloroacetylisocyanate to generate the initiator, shown in Scheme 28. The ATRP of MMA was carried out using the CuCl/dNbpy catalyst system at 130 °C for 30 h producing two different polymer samples that had M_n= 15,000 and 28,000 with M_w/M_n=1.18 and 1.14, respectively. The electropolymerization was carried out in the presence of two different supporting electrodes, p-

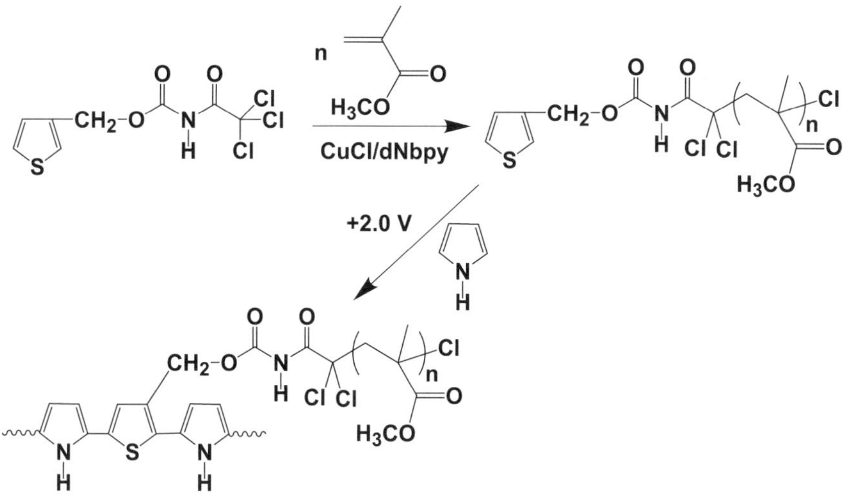

Scheme 28. ATRP of pMMA followed by electropolymerization of pyrrole to produce the block copolymers [251]

toluenesulfonic acid and sodium dodecylsulfate. Cyclic voltammetry indicated that the original pMMA films were not electroactive; however, after incorporation of the PY, there was a significant change in the electroactivity, resulting in an increasing redox peak with an increase in the number of scans [251]. Although FTIR analysis showed the presence of the pMMA in the block copolymer samples, after washing the films in acetonitrile, which will remove any homopolymer impurities of pMMA, scanning electron microscopy indicated that blocking efficiency was low (8–16%) [251].

3.2.2.2
CRP and Dendrimers/Hyperbranched Polymers

Frechet et al. reacted a halogen functionalized polyether dendron with HTEMPO to form a unimolecular initiator for TEMPO-mediated polymerization of St [252]. Several initiators were prepared, differing only in the number of generations present on the dendrimer. Well-defined dendritic-linear block copolymers were produced, with M_n=14,000 to 91,000 and M_w/M_n=1.14 to 1.42, depending on the macroinitiator used [252]. Prior work attempted to use a TEMPO-functional dendron as the polymerization mediator [253]; however, due to diffusion effects, the control over the polymerization was not as good as with these unimolecular initiators. Similarly, various benzyl halide (Br or Cl) terminated dendrimers were used for the ATRP of St in conjunction with a CuCl and either 4,4′-(bis-n-heptyl)-2,2′-bipyridyl or bpy as the ligand [252]. Again, the block copolymers were well-defined with predictable molecular weights and narrow molecular weight distributions. Interestingly, there was a single T_g for all the block copolymers, regardless of the generation of the dendrimer initiator, suggesting that the two blocks do not phase separate, (blends of the two polymers exhibit two T_gs) [252]. Expanding on this work, the authors used dendrons with exterior ethyl esters as the initiators for similar copolymerizations [254]. This provided a direct route to materials with carboxylic acids, butyl amides, and alcohols on the periphery via transformation chemistry, which were to be investigated for several uses, including surface modifiers and adhesives.

Later work by Frechet et al. attempted to prepare ABA diblock copolymers using the same methodology [255]. Two polyether dendrons were connected to a dihydroxy functional TEMPO moiety, which was then used to initiate, as well as control, the polymerization of St and formed "dumbbell-shaped" block copolymers (Scheme 29). The ABA block copolymers were formed; however, there was some initial contamination from mono- and bis-dendritic species, which were removed by column chromatography. In addition, ^{1}H NMR analysis determined that the purified ABA triblock copolymers were also contaminated with AB diblock copolymers. This was attributed to the persistent radical effect, as well as inherent problems with the mobility of the bulky dendron counter radical, both of which will result in the AB diblock copolymers. Thermal polymerization may also contribute to diblock formation, as the radicals generated will be trapped by the counter radical dendron, but would not contain the initiating dendron moiety. Pure ABA triblock copolymer was obtained after two purification cycles using column chromatography [255].

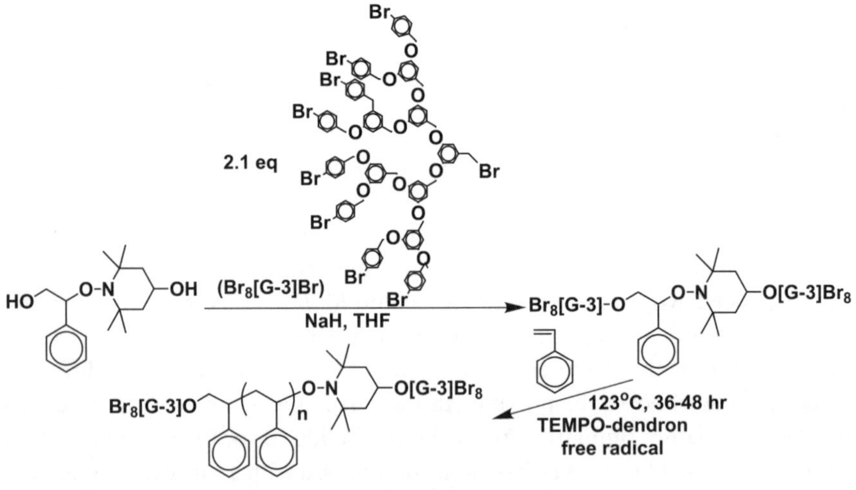

Scheme 29. Functionalization of a polyether dendron using a dihydroxy-TEMPO moiety followed by the formation of ABA "dumbbell-shaped" block copolymers with St [255]

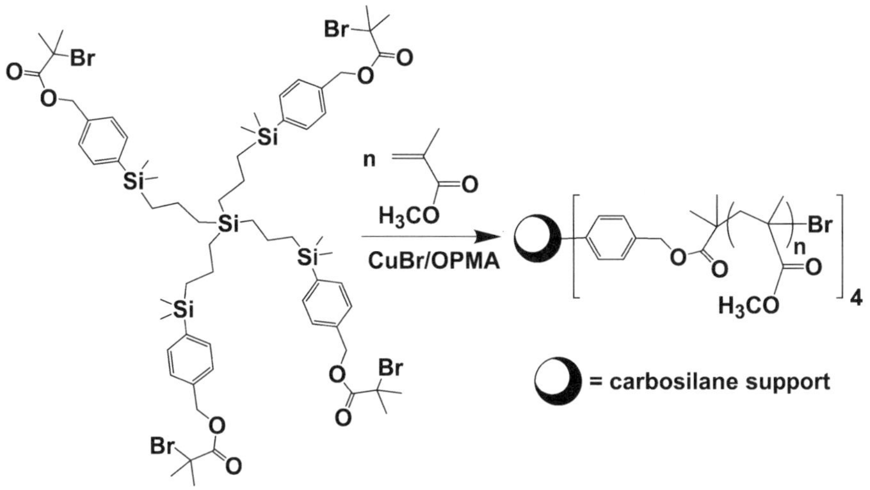

Scheme 30. 0-th generation carbosilane dendrimer used as a macroinitiator for the ATRP of MMA [256]

Carbosilane dendrimers were also used as macroinitiators for ATRP by van Koten et al. [256]. Using bromoisobutyryl functional 0-th and 1-st generation dendrimers as the initiators (Scheme 30), the ATRP of MMA was carried out using the CuBr/n-octyl-2-pyridylmethanimine (OPMA) catalyst system, resulting in block copolymers with M_n=13,900 (4 initiating sites) and 33,000 (12 initiating sites) and M_w/M_n=1.18 and 1.29, respectively [256]. There was significant dendrimer-dendrimer coupling in the polymerization using the 12-site initiator and this was present from the early stages of the reaction. Polymerizations performed using macroinitiators with varying degrees of functionality showed that the molecular weight evolution was similar for both the 6- and 12-site initiators, indicating that GPC analysis does not provide accurate molecular weight results due to the compact nature of the star shaped structure [256]. This has proven to be yet another approach to incorporate inorganic functionality into a block copolymer.

Hyperbranched polymers may be prepared by the self-condensing vinyl polymerization (SCVP) [257] of AB* star monomers by a controlled free radical process, such as ATRP [258]. The result, under certain conditions, is a highly branched, soluble polymer that contains one double bond and, in the absence of irreversible termination, a large quantity of halogen end groups equal to the degree of polymerization which can be further functionalized [87] (Fig.35). Two examples explored in detail by ATRP are vinyl benzyl chloride (VBC, p-chloromethylstyrene) [258] and 2-(2-bromopropionyloxy)ethyl acrylate (BPEA) [259–261] both depicted in Fig. 35. Several other (meth)acrylates with either 2-

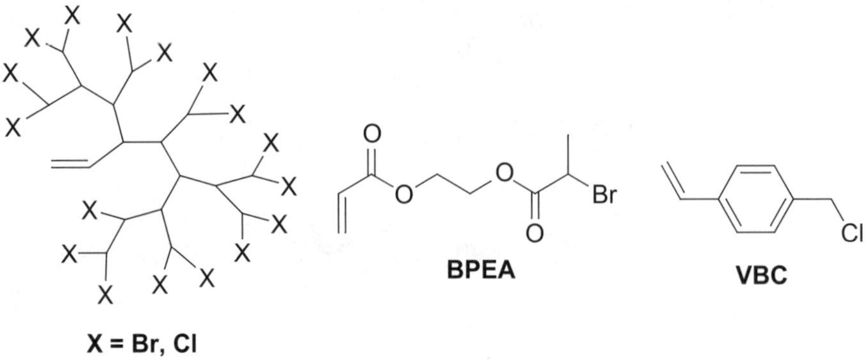

Fig. 35. Illustration of a hyperbranched polymer with halogen chain ends (X), 2-(2-bromopropionyloxy)ethyl acrylate (BPEA), and vinyl benzyl chloride (VBC)

bromopropionate or 2-bromoisobutyrate groups were also used [262]. It should be noted that, under certain conditions, linear homopolymers of the AB* monomers can be synthesized as intermediates toward other chain architectures [261].

Hyperbranched polymers synthesized by ATRP using "mixed monomers", structures that contain combinations of (meth)acrylates with α-haloesters [262], have also been reported. For example, 2-(2-bromopropionyloxy)ethyl methacrylate (BPEM) contains the methacrylate and bromopropionate groups which form tertiary and secondary radicals, respectively. Likewise, the monomer 2-(2-bromoisobutyryloxy)ethyl acrylate (BIEA) contains the secondary acrylate group with a tertiary 2-bromoisobutyrate fragment. With these monomers, highly branched macromolecules were obtained. In a similar way, macroinitiators were used to reduce the proportion of branched units [263]. They may be considered a segmented block copolymers.

Another approach to incorporate a block structure was to use the multifunctional precursor shown in Fig. 35 and grow blocks from the core of the hyperbranched structure. Such star-like polymers with ~80 pnBA blocks were obtained [130]. In a similar way, hyperbranched polymers from VBC were used to initiate the ATRP of nBA [130] and St [264]. Dendrigraft polystyrene was found to display a lower intrinsic viscosity and higher thermal stability than linear polystyrenes [264].

Hydrophilic pEG or pentaerythritol ethoxylate cores with hyperbranched polystyrene arms were prepared by reacting PEG or pentaerythritol ethoxylate with 2-bromopropionyl bromide followed by the ATRP of the macroinitiator and chloromethylstyrene to produce the amphiphilic hyperbranched polymer. Depending on the functionality of the macroinitiator, the products have either a dumbbell or 4-arm starburst structure. The dumbbell polymers tend to have higher molecular weights, while the starburst polymers have rather low molecular weights [265].

Heat-resistant hyperbranched copolymers of VBC and CMI have also been synthesized by ATRP. Under identical polymerization conditions and after the same reaction time, high monomer conversions occurred near the equimolar feed composition, indicating the formation of charge transfer complexes between VBC (electron-donor) and maleimide (electron-acceptor). As expected, the T_g of the copolymer increased with an increasing content of maleimide in the feed [266].

Frey et al. have synthesized hyperbranched polyglycerols (pG) using a technique termed "ring-opening multibranching polymerization" [267]. They have combined this technique with ATRP to prepare multi-arm star polymers of pG with MA [268]. A pG containing an average of 64 hydroxyl groups was transformed into an ATRP initiator via an esterification reaction with 2-bromoisobutyryl bromide. The extent of functionalization was held to <100% to prevent any remaining acid from participating in unwanted side reactions. Chain extension of pG containing 71 and 86% bromine functionalities with MA was carried out using the CuBr/PMDETA catalyst system. The rate of polymerization was a function of the ratio of monomer:initiating sites, with higher ratios producing slower rates. The experimental molecular weights increased linearly with conversion and the final M_w/M_n were <2.5. Gelation was observed at >35% monomer conversion, indicating that extensive crosslinking occurred. At lower monomer:initiator ratios, it was necessary to decrease the concentration of catalyst to achieve monomer conversions >30% [268]. However, the rate of the polymerizations was fast (38% after 25 min at monomer: initiator ratios of 116:1). No attempts to control the rate of reaction through addition of deactivator were attempted. This would have decreased the rate of polymerization and potentially allowed for higher monomer conversions.

3.2.2.3
CRP and Cationic Polymerizations

3.2.2.3.1
Vinyl Monomers

A pSt-Cl prepared using living carbocationic polymerization methods was used directly for the synthesis of block copolymers with St, MMA, or MA using ATRP for preparation of the second block by Coca and Matyjaszewski [222]. After isolating the macroinitiator (M_n=2080, M_w/M_n=1.17), the ATRP of St, MMA or MA was carried out using the homogeneous CuCl/dNbpy catalyst system at 100 °C (Scheme 31). The molecular weight of the macroinitiator increased with increasing monomer conversion and the molecular weight distributions were narrow. Block copolymers were also prepared directly by terminating the polymer-

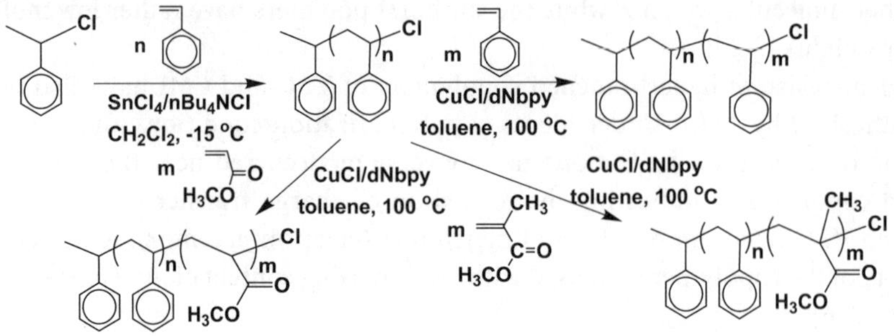

Scheme 31. The cationic polymerization of St followed by the ATRP of MMA and MA with a homogeneous CuCl/dNbpy catalyst system to produce AB block copolymers [222]

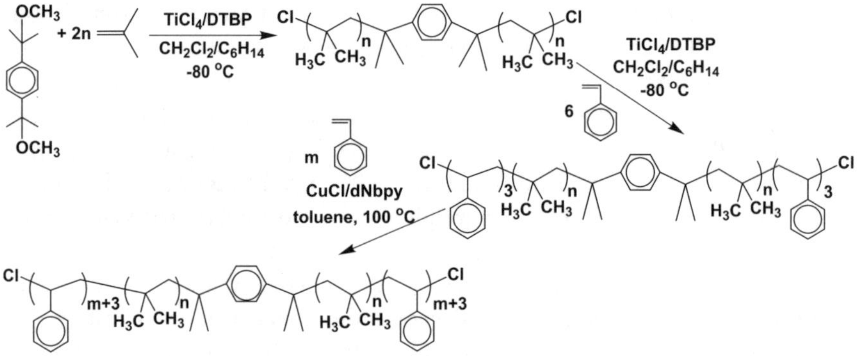

Scheme 32. Transformation of α,ω-difunctional pIB into a chloro-terminated macroinitiator followed by the ATRP of St [269]

ization by adding the second monomer (i.e., MA), removing the Lewis acid and solvent, then adding the ATRP catalyst solution. The results were nearly identical to the isolation technique, but provided block polymers with an even narrower molecular weight distribution (M_w/M_n=1.21 vs 1.57) [222].

Carbocationically prepared α, ω-difunctional polyisobutylene (pIB), possessing a terminal chlorine functionality was capped with several St units, then used as the macroinitiator for the ATRP of St, MMA, MA, and isobornyl acrylate (IA) using the CuCl/dNbpy catalyst system, as reported by Coca and Matyjaszewski and shown in Scheme 32 [269]. The molecular weights of the block copolymers increased significantly and although the molecular weight distributions broadened slightly, there was little evidence of unreacted macroinitiator in the GPC traces. The authors attributed this behavior to a slow cross-propagation step [269]. DSC analysis showed the presence of two T_gs corresponding to pIB (–71 °C) and pSt (91 °C), pMMA (94 °C), or pIA (93 °C) blocks. Ivan et al. also produced a macroinitiator via the same mechanism [270], which was subsequently used to initiate ATRP of St; however, their attempts to incorporate acetoxystyrene resulted in incomplete initiation, which the authors attributed to the poor solubility of the pIB in the reaction mixture [270].

3.2.2.3.2
Ring Opening Polymerizations

Poly(tetrahydrofuran), pTHF, prepared through living cationic ring-opening polymerization (CROP), was reacted with sodium OTEMPO to produce a counter radical for polymerization of St to form block copolymers, by Yoshida and Sugita [271]. In the presence of pTHF, the polymerization of St initiated by BPO was controlled, resulting in block copolymers with predictable molecular weights and narrow molecular weight distributions (M_w/M_n=1.22–1.40). The GPC traces showed complete consumption of the pTHF macroinitiator to form the desired block copolymer [271]. Later work used a difunctional initiator for the cationic polymerization, which was followed by reaction with the sodium OTEMPO, finally resulting in an ABA triblock copolymer with exterior pSt blocks (Scheme 33) [272].

In addition to using the pTHF as a macro-counter radical, it can be modified to produce a unimolecular polymeric initiator, as shown by Yagci et al. [273]. THF polymerization was initiated by a diazooxocarbenium ion, resulting in a polymer that, when heated, produced radicals that could be trapped by TEMPO to produce the desired macroinitiator. Heating St to 125 °C in the presence of these initiators gave block copolymers with higher molecular weights and relatively low polydispersities [273].

Cationically prepared pTHF can also contain a functional group capable of initiating an ATRP reaction, as shown by Kajiwara and Matyjaszewski [274]. Silver triflate in conjunction with 2-bromopropionyl bromide was used to initiate the ring-opening polymerization (ROP) of THF which, after quenching the re-

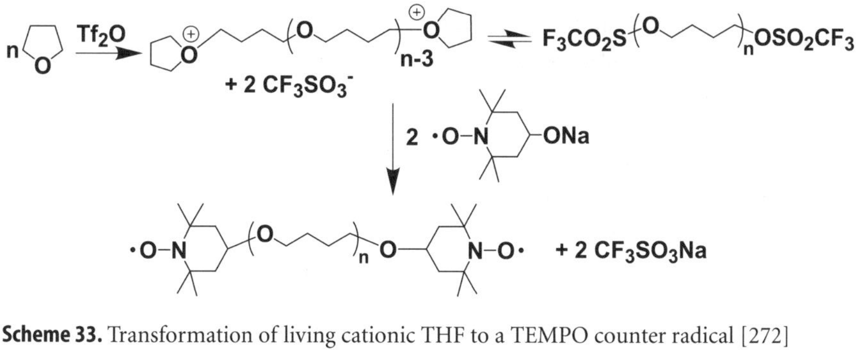

Scheme 33. Transformation of living cationic THF to a TEMPO counter radical [272]

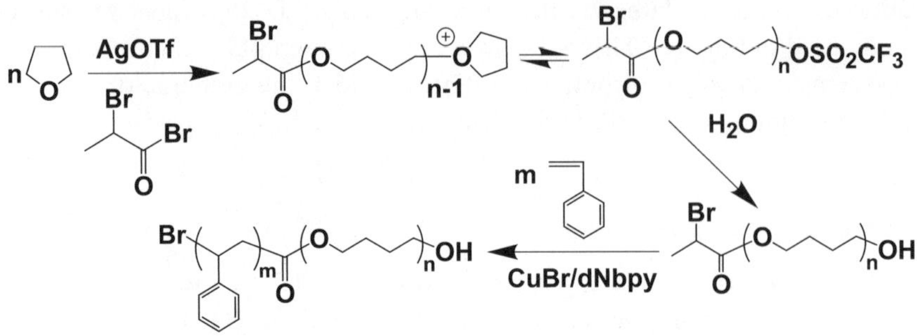

Scheme 34. Transformation of the CROP of THF to the ATRP of St [274]

action, allowed preparation of block copolymers from the bromine-containing chain end via ATRP (Scheme 34).

Block copolymers were synthesized with St, MA, and MMA using the homogeneous CuBr/dNbpy catalyst system. The molecular weight of the macroinitiator (M_n=15,400, M_w/M_n=1.39) increased with the formation of the second block (M_n=24,900 to 35,000) and the molecular weight distribution remained narrow (M_w/M_n>1.5) [274]. Difunctional macroinitiators were also prepared. Once the THF polymerization was complete, the polymer was reacted with sodium 2-bromopropionate to produce the ATRP initiator, which was then used for the preparation of block copolymers, again with St, MMA, and MA. For both St and MA, the polymerization leading to the preparation of the second block was incomplete; however, it proceeded smoothly for MMA. In both systems, block formation was confirmed through DSC analysis [274].

The reverse order for the formation of the blocks has also been utilized. Ying et al. synthesized block copolymers containing pTHF and p(2-methyloxazoline) (MO); however, they first used ATRP for the St polymerization, then chain extended the pSt macroinitiator by the ROP of THF or MO via a cationic mechanism [275-277]. Chloro-terminated pSt (M_n=4100–15,150, M_w/M_n=1.34–1.70) was reacted with a THF solution containing $AgClO_4$ to form the cationic species, then the polymerization of THF was allowed to proceed for a given time before it was quenched by addition of a protonic substance (water, alcohol). The procedure was similar for the MO chain extension. Although both the experimental and characterization details were minimal for the THF polymerization, the results presented indicated that more efficient blocking could be achieved with use of lower molecular weight macroinitiators [276]. The block copolymers prepared with MO appeared to be well-defined based on the molecular weight results presented; however, the molecular weight distributions were significantly broader when the chlorine-terminated pSt was used than when the pSt-Br (M_w/M_n>1.35 vs <1.2) macroinitiator was employed [276]. This suggests that the

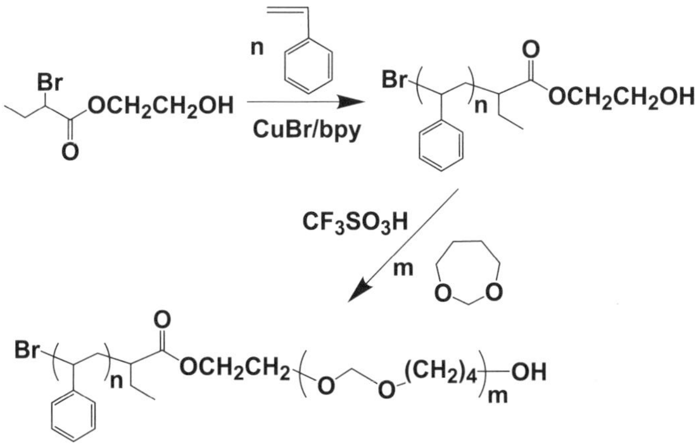

Scheme 35. ATRP of St followed by cationic ROP of 1,3-dioxepane [280]

transformation reaction is either faster or more complete in the presence of the bromine-terminated macroinitiator, as would be expected for a nucleophilic substitution reaction [278].

Xu and Pan reported on a similar system. Bromine-terminated pSt was reacted with silver perchlorate to prepare a macroinitiator for the living CROP of THF [279]. Using macroinitiators of M_n=4250 and 7890, block copolymers were synthesized with M_n=9420 to 17,450 and M_n=11,860 to 32,980, respectively. The M_w/M_n were <1.45 in all cases. They also prepared a pSt based macroinitiator via ATRP using a hydroxy functional bromine-based initiator which was then used as a chain transfer agent for the CROP of 1,3-dioxepane (DOP, Scheme 35) [280]. The pSt macroinitiators obtained using the CuBr/bpy catalyst system had M_n= 6300 (M_w/M_n=1.18) and 11,360 (M_w/M_n=1.19). The ROP of DOP, using a slow addition of monomer technique, resulted in block copolymers with M_n=17,540 (M_w/M_n=1.43) and M_n=30,880 (M_w/M_n=1.46), as determined by ^{1}H NMR. Control over the molecular weights and the molecular weight distributions was lost if higher molecular weight blocks of pDOP were targeted [280].

In an interesting experiment, Yagci et al. polymerized cyclohexene oxide (CHO) via a photosensitized cationic polymerization with an initiator that contained a TEMPO moiety capable of CRP [281]. Anthracene was reacted with N-ethoxy-2-methyl pyridinium hexafluorophosphate, which produced a radical cation that could then be trapped with TEMPO to create the dual initiating species capable of both cationic and nitroxide-mediated polymerizations (Scheme 36).

After the polymerization of CHO, the macroinitiator was purified and used for the polymerization of St. The resulting block copolymers had increased mo-

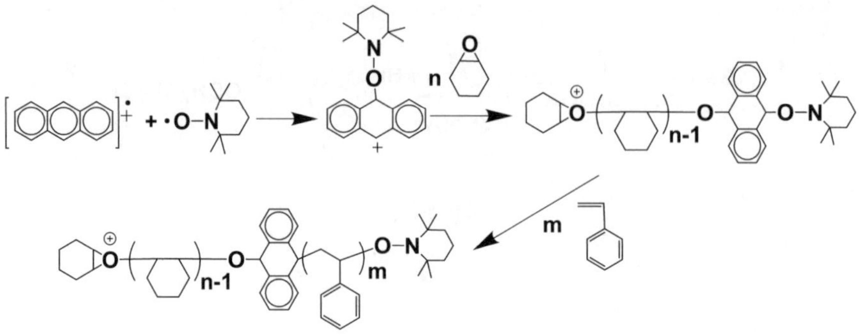

Scheme 36. Sensitized cationic ROP of cyclohexene oxide followed by the TEMPO-mediated CRP of St [281]

lecular weights (M_n=3200 to 40,000) with no increase in the molecular weight distribution (M_w/M_n=1.5) [281]. The GPC traces showed high blocking efficiency and the IR analysis indicated the presence of two different polymer segments in the samples.

3.2.2.4
CRP and Anionic Polymerizations

3.2.2.4.1
Vinyl Monomers

Living anionic polymerization has also been used to prepare macroinitiators for subsequent chain extension to form block copolymers using CRP methods. There are several examples in the literature of using anionically prepared pBD to initiate either ATRP or nitroxide-meditated polymerizations. Priddy et al. used an epoxy-functional TEMPO moiety to terminate the living anionic polymerization of BD, which was then employed as an initiator for the polymerization with St to create high-impact pSt in situ [282]. Based on ^{1}H NMR analysis, the pBD had a chain end functionality >95%, resulting in a small amount of tailing in the GPC traces of the block copolymer; however, the molecular weight distribution remained narrow (M_w/M_n=1.22). After degradation of the block copolymer, the pSt segment had an M_n=12,000 with M_w/M_n=1.18, in good agreement with the original estimates from analysis of the block copolymer (M_n=15,000 via GPC or 11,400 via ^{1}H NMR) [282].

Miura et al. terminated a living anionic polymerization of BD with several different types of nitroxides to prepare CRP initiators [283]. The bulk polymerization of St initiated by four different nitroxide moieties proceeded with a linear increase of the molecular weight with monomer conversion and molecular weight distributions <1.2. To synthesize the macroinitiators for preparation of

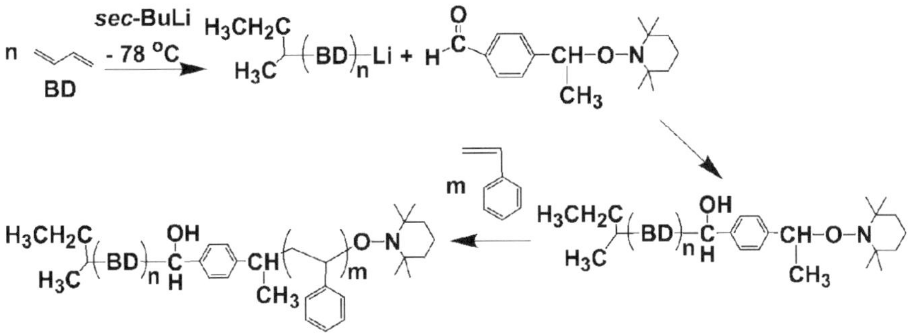

3 Linear Block Copolymers

Scheme 37. Anionic polymerization of BD followed by transformation to a TEMPO-based macroinitiator and the CRP of St [283]

block copolymers, pBD was end-capped with a formyl-functional TEMPO derivative (Scheme 37).

^{1}H NMR and vapor pressure osmometry analysis gave M_n=5800 of the pBD, and confirmed that the chain ends were fully functionalized with the TEMPO. However, the GPC traces of a block copolymer formed with St showed some evidence of unreacted macroinitiator as well as a high molecular weight shoulder associated with coupling products [283].

Acar and Matyjaszewski reported on transformation of anionically prepared pSt to an ATRP initiator which was chain extended with various other monomers to create AB diblock copolymers [284]. The polystyryl anion was reacted with 2-bromoisobutyryl bromide to produce the macroinitiator. Initially, the molecular weight distribution of the pSt post-functionalization was bimodal as a result of coupling between the anion and the bromine-containing chain end, so the polystyryl anion was first reacted with styrene oxide (SO), then with 2-bromoisobutyryl bromide to generate a macroinitiator with a monomodal molecular weight distribution [284]. Chain extension of the pSt-Br with St, MMA, MA, nBA, and a mixture of St and AN proceeded with linear first-order kinetics as well as a linear increase in molecular weight with increasing conversion, with narrow molecular weight distributions (M_w/M_n<1.2). The only ATRP chain extension reaction that resulted in a block copolymer with a broader molecular weight distribution was chain extension with MMA (M_w/M_n=1.36). Halogen exchange [175] was not utilized; therefore, the rate of polymerization for the MMA was faster than the rate of crosspropagation and the polymerization was less controlled [284]. In addition to the AB diblock copolymers, ABA triblock copolymers were prepared. Living pSt was chain extended with IP, which was then capped with SO, followed by transformation into the ATRP initiator. Chain extension with St produced unsymmetrical ABA triblock copolymers and although the reaction proceeded with little termination and a linear increase of molecular weight with conversion, the GPC traces had a higher molecular weight shoulder; this was attributed to a "grafting through" side reaction with the unsaturated pIP segments in the backbone [284].

Ying et al. have reported on similar work [285–289]. The authors first anionically polymerized St, then end-capped the living chain end with ethylene oxide, followed by reaction with trichloroacetyl chloride (TCAC) to form the ATRP initiator. In some instances, BD was added prior to the transformation reaction to create an AB diblock copolymer, which was then chain extended. The acylation efficiency was high when an excess of TCAC was used and the reaction temperature was raised to 60 °C [287]. In most chain extensions, with St or BA, complete consumption of the macroinitiator was not achieved and bimodality was observed in the GPC traces. However, chain extension of the pSt-Cl$_3$ macroinitiator with MMA was carried out successfully using the CuCl/bpy catalyst system at 15 °C in 50% xylene solution to generate a copolymer with M_n=31,500 and M_w/M_n=1.39 [287]. In related work, bromine-terminated pSt was also prepared; however, this was accomplished through the bromination of an α-methyl styrene (MeSt) capped polymer chain, as opposed to the direct incorporation of an active alkyl halide moiety [286, 288]. Varying the reaction conditions had little effect on the efficiency of bromination; a slight excess of Br$_2$ was sufficient to achieve >90% efficiency, regardless of the molecular weight of the pSt, the

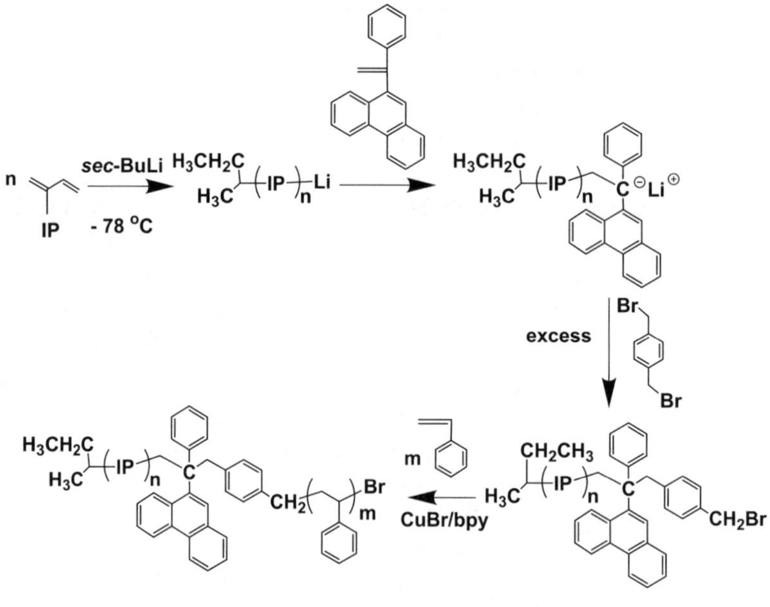

Scheme 38. Anionic polymerization of IP end-capped with a fluorescent probe, followed by transformation and the ATRP of St [238]

amount of MeSt present, or the amount of solvent used. The molecular weights of the block copolymer with MMA increased linearly with conversion and the molecular weight distributions were narrow. ^{1}H NMR analysis indicated the presence of both pSt and pMMA blocks [288].

Winnik et al. used living anionic polymerization to prepare pIP, which was then reacted with a fluorescent dye derivative, followed by a transformation to an ATRP initiator through incorporation of α, α-dibromoxylene (DBX, Scheme 38) [238]. The fluorescent dye served two purposes: 1) to decrease the amount of chain coupling that can occur once the DBX is added and 2) to measure the distance between two polymer chains based on the donor-acceptor relationship between the dyes. Block copolymers with St were successfully prepared using the pIP capped with either the donor or acceptor dye derivative. Although coupling occurred and produced dead macroinitiator, the contaminant could be removed through extraction, leaving the well-defined block copolymers [238].

3.2.2.4.2
Ring Opening Polymerizations

ROP can also be carried out through an anionic mechanism. Yoshida and Osagawa reacted an excess of HTEMPO with triethylaluminum to form an aluminum tri (OTEMPO) complex capable of initiating the ROP of ε-caprolactone (CL) [290]. The pCL was then used as a counter radical for the polymerization of St initiated by BPO. There was good agreement between the molecular weight calculated based on the ratio of intensities of the terminal-hydroxy protons to the methylene protons from ^{1}H NMR and that based on the content of TEMPO per gram of polymer from UV analysis, indicating that TEMPO had been incorporated into the chain ends. Chain extension with St was successful, yielding block copolymers with M_n=19,900 to 26,200 with M_w/M_n=1.37 to 1.42 [290]. A detailed investigation showed that the polymerization proceeded with a minimal amount of termination and that molecular weights were predictable based on concentration of pCL-TEMPO macroinitiator. DSC analysis confirmed the presence of two distinct blocks, with T_gs=–56 °C (pCL) and 90 °C (pSt) [290].

Ethylene oxide (EO) can also be polymerized via ROP. Huang et al. used potassium 2-dimethylaminoethoxide as the initiator for the polymerization, followed by quenching with methanol. The dimethylamino-capped polymer was exposed to 365-nm light in the presence of benzophenone and capped with HTEMPO to form the initiator (Scheme 39) [291].

The capping efficiency ranged from 85 to 89%. When used as an initiator, the molecular weights of the resulting block copolymers increased with conversion and the molecular weight distributions remained narrow (M_w/M_n<1.5). Contamination with homopolymer of St, generated through the thermal process, was only found when the concentration of St was increased to such an extent that

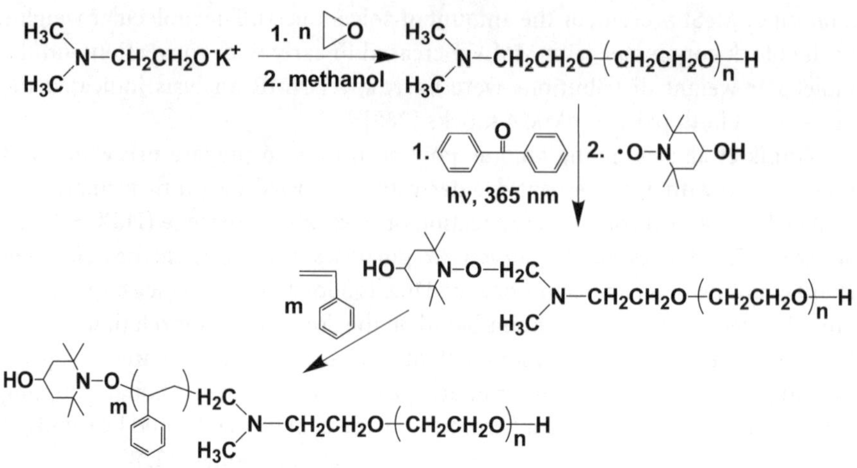

Scheme 39. ROP of ethylene oxide followed by transformation to a TEMPO-terminated macroinitiator and the CRP of St [291]

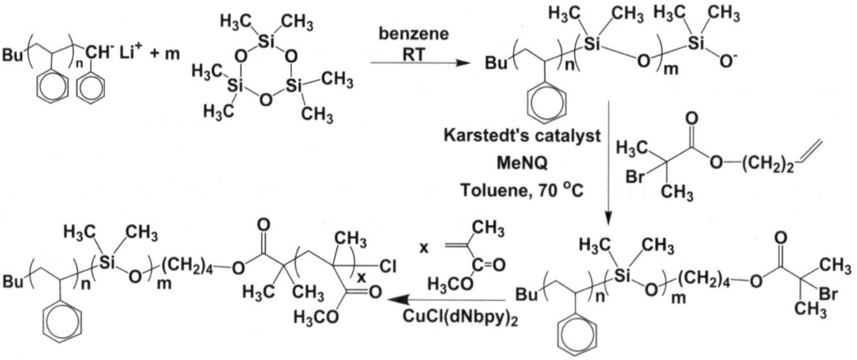

Scheme 40. The anionic polymerization of St followed by the ROP of D_3, transformation to an ATRP initiator and the CRP of MMA [234]

there was not sufficient free HTEMPO to control the polymerization. DSC analysis showed the presence of T_gs=–19 °C (pEO) and 98 °C (pSt), indicating the presence of the two blocks in the copolymer [291].

Siloxanes are another class of cyclic monomers that have been incorporated into CRP copolymers via a transformation reaction. Miller and Matyjaszewski used n-butyllithium to initiate the ROP of hexamethylcyclotrisiloxane (D_3), then terminated the polymerization with chlorodimethylsilane [234]. Allyl 2-bromoisobutyrate was then incorporated via a hydrosilylation reaction using Karstedt's catalyst. This reaction did not achieve a high end-functionality, but use of 3-butenyl 2-bromoisobutyrate, and the addition of 0.2 mol% of 2-methyl-1,4-

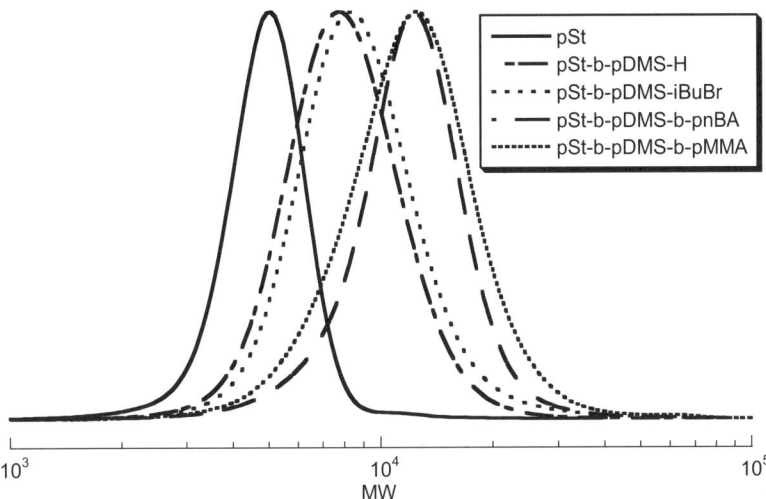

Fig. 36. GPC chromatograms of ABC triblock copolymers of pSt-b-pDMS-b-pMMA and pSt-b-pDMS-b-pnBA using hydrosilation and ATRP. Reprinted with permission from [234]. Copyright (1999) American Chemical Society.

naphthoquinone relative to the concentration of pDMS, resulted in a functionality of >0.9 [234]. Use of this macroinitiator for the polymerization of nBA proceeded with little termination and with a linear increase in the molecular weight with monomer conversion. The polydispersities were $M_w/M_n < 1.2$. The GPC traces also indicated that the chain end functionality was maintained throughout the polymerization. ABC triblock copolymers were also prepared. Living anionic pSt was chain extended with D_3, followed by incorporation of the ATRP initiator. The ATRP of both nBA and MMA was carried out and the molecular weight of the macroinitiator increased from M_n=7900 to M_n=10,200 and 10,100, respectively, as shown in Scheme 40 and Fig. 36. The molecular weight distribution remained narrow [234].

3.2.2.5
CRP and ROMP

Matyjaszewski et al. demonstrated that living ring opening metathesis polymerization (ROMP) could also be combined with ATRP to produce novel block copolymers [292]. ROMP of norbornene (NB) and dicyclopentadiene (CPD) were performed using an Mo-alkylidene complex, followed by reaction with *p*-(bromomethyl) benzaldehyde to generate a benzyl bromide terminated polymer capable of being used as a macroinitiator for ATRP (Scheme 41).

The homogeneous ATRP of St and MA were carried out using the CuBr/dNbpy catalyst system and the macroinitiators [292]. The block copolymers, pNB-*b*-

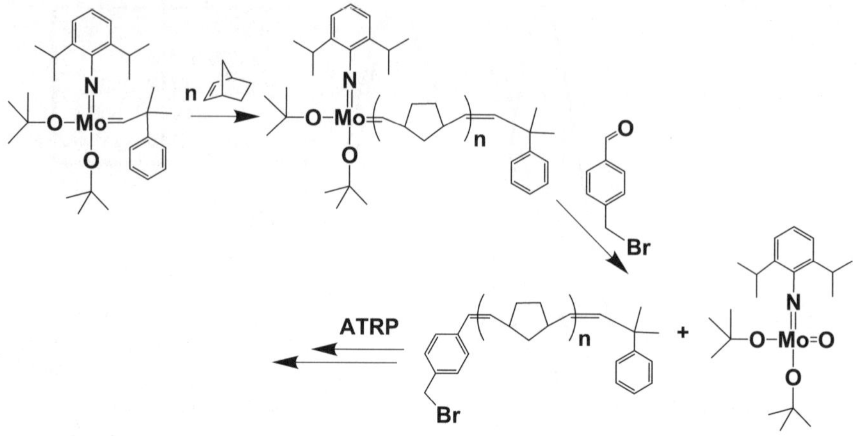

Scheme 41. ROMP of norbornene followed by transformation to an ATRP macroinitiator [292]

pSt, pNB-*b*-pMA, pCPD-*b*-pSt, and pCPD-*b*-pMA were well-defined with predictable molecular weights and narrow molecular weight distributions, M_w/M_n <1.4. The GPC traces showed a clean shift of the distributions to higher molecular weights, indicating good chain end functionality [292]. For example, polymerization of styrene and methyl acrylate from a PNB macroinitiator (M_n= 30,500, M_w/M_n=1.09) yielded pNB-*b*-pSt with M_n=110,400, M_w/M_n=1.06 (cf. Fig. 37) and pNB-*b*-pMA with M_n=85,100, M_w/M_n=1.07. In all of the polymerizations two glass transition temperatures were observed, indicating microphase separation of the two block segments.

Grubbs et al. prepared telechelic pBD through ROMP of cyclooctadiene in the presence of a chain transfer agent that impartedd either allyl chloride or 2-bromopropionate groups onto the chain ends, forming polymers that were capable of initiating ATRP and producing ABA triblock copolymers [293]. ^{1}H and ^{13}C NMR characterization showed the pBD had a perfect 1,4-microstructure. In the presence of a CuCl/bpy catalyst, the pBD (M_n=2400, M_w/M_n=1.59) was used as a macroinitiator for the ATRP of St. The molecular weights of the polymers were predictable (M_n=4800–13,800) with relatively narrow molecular weight distributions (M_w/M_n<1.55). Narrower molecular weight distributions were obtained using catalysts based on the more soluble 4,4′-diheptyl-2,2′-bipyridine ligand (M_w/M_n=1.25) [293]. The ATRP reaction was also carried out using a one-pot technique. After quenching the ROMP with ethyl vinyl ether, St, phenyl ether, bpy, Cu(0) [242] and CuBr$_2$ were added to the flask, which was heated to 130 °C for 7 h. The resulting ABA block copolymer had M_n=8000 with M_w/M_n= 1.63. While the Cl-based system was useful for the ATRP of St, the ATRP of MMA was not as successful. The GPC traces were bimodal, which resulted from the

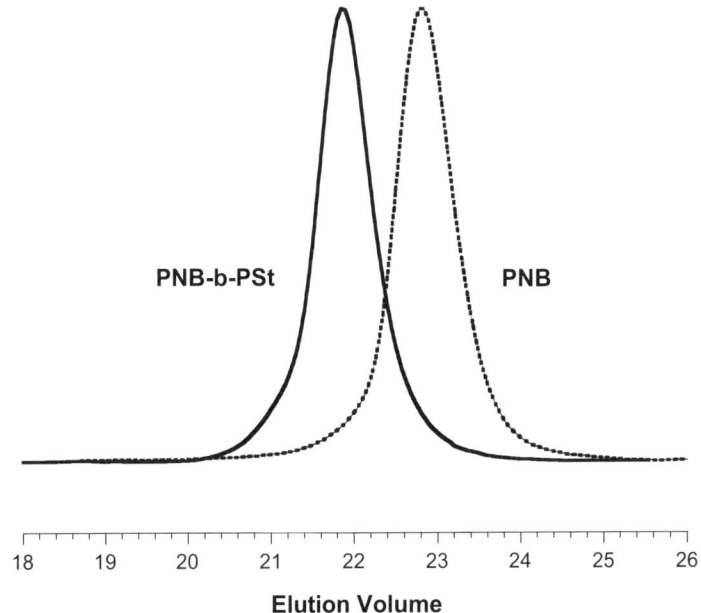

Fig. 37. GPC traces for pNB-b-pSt (M_n=110,400, M_w/M_n=1.06) prepared via a transformation of ROMP to ATRP. Reprinted with permission from [292]. Copyright (1997) American Chemical Society.

mismatch between the rate constants of cross-propagation and polymerization. Using a bromine-terminated macroinitiator (M_n=5500, M_w/M_n=1.58) in the presence of the CuCl/bpy catalyst to invoke halogen exchange [175], on the other hand, produced well-defined block copolymers with M_n=4700 to 41700 and M_w/M_n<1.7.

3.2.3
Summary

Numerous examples exist of combining CRP methods with other polymerization techniques for preparation of block copolymers. Non-living polymerization methods like condensation, free-radical, and redox processes can easily be combined with CRP to produce novel materials. Transformation chemistry may be the only route to incorporate polymers like polysulfones (as described above), polyesters, or polyamides that are prepared solely through condensation processes into subsequent CRP to form block copolymers with vinyl monomers. The same can be said of polymers prepared through coupling techniques, like poly(phenylenevinylene) and poly(methylphenylsilylene), which can maintain their conductive or photoluminescence properties, but become easier to process

Table 8. Summary of block copolymers from a combination of non-living and CRP polymerization techniques

Methods	Monomers	Comments	Investigator
Condensation/ATRP	p(Sulfone)/St or nBA	Two T_gs present	Gaynor and Matyjaszewski [223]
Coupling/ATRP	MPSi/St	Mixture of AB and ABA chains	Jones et al. [240]
Coupling/ATRP	Oligophenylenes/St	M_n=7900–20,600, M_w/M_n=1.17–2.15	Kallitsis et al. [241]
Condensation/nitroxide	PV/St	M_n=9000	Hadziioannou et al. [243]
Radical/nitroxide	IP/St	ABA block copolymers	Priddy et al. [244]
Radical/nitroxide	PDMS/St, then AcOSt	M_n=135,000, M_w/M_n=1.43	Yoshida and Tanimoto [245]
Radical/ATRP	VA/St	M_n=91,600, M_w/M_n=1.80	Matyjaszewski et al. [246]
ATRP/radical	nBA/VA	Low blocking efficiency only about 50%	Matyjaszewski et al. [246]
Redox/ATRP	VA/St	M_n=24,300, M_w/M_n=1.42	Matyjaszewski et al. [246]
ATRP/redox	nBA/VA	M_n=4450, M_w/M_n=2.58	Matyjaszewski et al. [246]
Telomerization/ATRP	VA/St	M_n~8000, M_w/M_n<1.5	Boutevin et al. [247]
Radical/ATRP	nBA/St	M_n=22,620, M_w/M_n=1.66	Destarac and Boutevin [248]
Telomerization/ATRP	VDF/St	M_n=37,360, M_w/M_n=1.65	Ying et al. [249]
ATRP/electropolym	MMA/pyrrole	M_n=15000–28000 M_w/M_n=1.18–1.14	Yagci et al. [251]
Dendrimer/nitroxide	Polyether/St	M_n=14,000–91,000 and M_w/M_n=1.14–1.42	Frechet et al. [252]
Dendrimer/ATRP	Polyether/St	Single T_g, two blocks miscible	Frechet et al. [252]
Dendrimer/nitroxide	Polyether/St	ABA "dumbbell", not pure ABA	Frechet et al. [255]
Dendrimer/ATRP	Carbosilane/MMA	4–12 Sites, M_n= 13,900–33,000, M_w/M_n=1.18–1.29	Haddleton et al. [256]
Hyperbranched/ATRP	Glycerol/MA	71–86% ATRP functional, M_w/M_n<2.5	Frey et al. [268]

as part of a well-defined block copolymer. Table 8 summarizes the above discussion section on this topic.

CRP methods may also be combined with ionic polymerization methods and the polymers may not require any type of transformation chemistry before their use as macroinitiators. For example, pSt prepared cationically and having chlo-

3 Linear Block Copolymers

Table 9. Block copolymers prepared from a combination of ionic (+=cationic, –=anionic) and CRP polymerization techniques

Methods	Monomers	Comments	Investigator
+/ATRP	St/St, MMA, MA	M_n=5080–11,090, M_w/M_n=1.10–1.57	Coca and Matyjaszewski [222]
+/ATRP	IB/St, MMA, MA, IA	difunctional, M_n=28,800–33,500, M_w/M_n=1.14–1.47	Coca and Matyjaszewski [269]
+ ROP/nitroxide	THF/St	AB and ABA (counter radical), M_w/M_n=1.22–1.40	Yoshida and Sugita [271, 272]
+ ROP/nitroxide	THF/St	Unimolecular initiator	Yagci et al. [273]
+ ROP/ATRP	THF/St, MA, MMA	AB and ABA, M_n=24900 – 35000, M_w/M_n>1.5	Kajiwara and Matyjaszewski [274]
ATRP/+ ROP	St/THF, MO	Lower M_n pSt gave higher blocking eff.	Ying et al. [277]
ATRP/+ ROP	St/THF	M_n=9420–32,980, M_w/M_n<1.45	Xu and Pan [279]
ATRP/+ ROP	St/DOP	M_n=17,540 (M_w/M_n=1.43) and M_n= 30,880 (M_w/M_n=1.46)	Pan et al. [280]
Photosensitized+ nitroxide	/CHO/St	M_n=3200 to 40,000, M_w/M_n=1.5	Yagci et al. [281]
–/nitroxide	BD/St	M_w/M_n, block=1.22, M_n pSt=12,000, M_w/M_n=1.18	Priddy et al. [282]
–/nitroxide	BD/St	Inefficient blocking	Miura et al. [283]
–/ATRP	St/St, MMA, MA, nBA	M_w/M_n<1.2	Acar and Matyjaszewski [284]
–/ATRP	St-IP/St	ABA, side reaction with unsaturations	Acar and Matyjaszewski [284]
–/ATRP	St/St, nBA, MMA	Chlorine chain ends, only MMA efficient	Ying et al. [289]
–/ATRP	IP/St	Fluorescent dye junction	Winnik et al. [238]
–ROP/nitroxide	CL/St	Counter radical, M_n=19,900–26,200, M_w/M_n=1.37–1.42	Yoshida and Osagawa [290]
–ROP/nitroxide	EO/St	Capping efficiency <90%, M_w/M_n <1.5	Huang et al. [291]
–ROP/ATRP	St then PDMS/ nBA, MMA	M_n=10,200 and 10,100	Miller and Matyjaszewski [234]
ROMP/ATRP	CPD, NB/St, MA	M_w/M_n<1.4	Matyjaszewski et al. [292]
ROMP/ATRP	BD/St	ABA blocks, M_n=4800–13800, M_w/M_n<1.55	Grubbs et al. [293]

rine chain ends was used for the ATRP of various monomers without the need for any transformation reaction (Table 9, entry 1) pIB, a polymer that has a very low T_g, can also be polymerized cationically, end-capped, and directly used for ATRP to prepare thermoplastic elastomers when combined with monomers that form higher T_g polymers (Table 9, entry 2). Both cationic and anionic methods have been used extensively for ring-opening polymerizations, which, when terminated with a protic substance, generally results in polymers with hydroxy chain ends that are easily modified into CRP initiators/counter radicals. This has been demonstrated with polymers prepared from THF, ethylene oxide, oxazolines, caprolactone, and dimethylsiloxane, as detailed in Table 9. While diene-based monomers prepared using free radical chemistry can be modified to form macroinitiators, the polymers are ill-defined. Well-defined pBD has been prepared by anionic methods, which was then transformed into macroinitiators for CRP. ROMP has also been combined with CRP methods to produce well-defined block copolymers. Table 9 contains details about the synthesis of all these types of block copolymers.

4
Other Chain Architectures

4.1
Graft Copolymers

There are three general methods for preparing graft copolymers: grafting onto, grafting through, and grafting from. Grafting onto requires the presence of complimentary functionalities on the graft unit and the backbone. Grafting through utilizes macromonomers, which are polymer chains that contain a copolymerizable moiety at the chain end. Homo- or copolymerization with another monomer produces the graft copolymer. The grafting from technique employs a backbone containing reactive sites that are capable of initiating a polymerization. Each method suffers from its own particular disadvantages, but steric hindrance of the reactive center is common to all the graft copolymer routes, affecting the grafting efficiency. Figure 38 illustrates the three approaches. The least utilized in CRP is perhaps the grafting onto approach, while grafting through and grafting from are common. These will be detailed below.

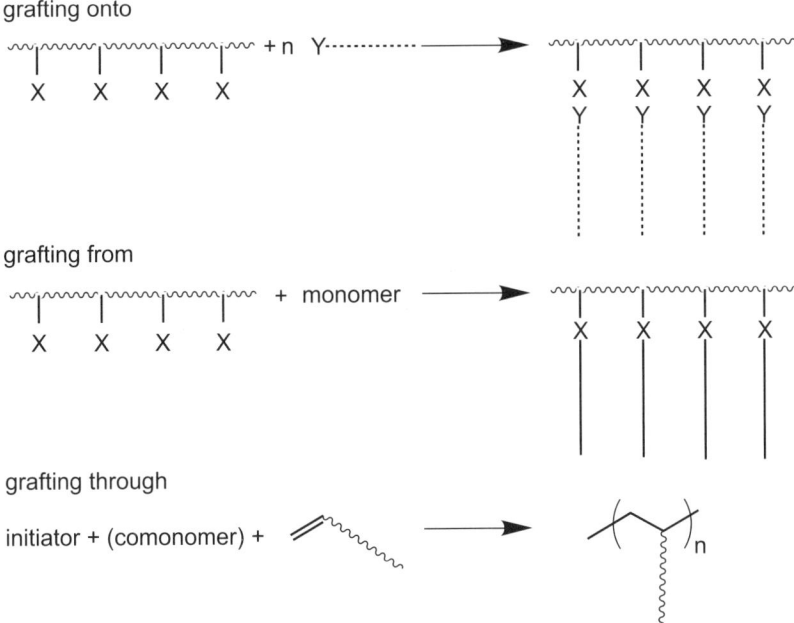

Fig. 38. Illustration of different methods of preparing graft copolymers

4.1.1
Grafting From

After establishing that a unimolecular TEMPO-based initiator successfully controlled the polymerization of St [62], Hawker et al. used the grafting from approach to demonstrate that graft copolymers were possible using CRP techniques [294]. The multifunctional macroinitiator was prepared by copolymerizing St with a vinyl benzyl TEMPO derivative free radically using AIBN at 60 °C. The macroinitiator was heated to 130 °C in the presence of more St, which activated the TEMPO bond, which led to the preparation of a graft copolymer. The molecular weight of the polymer increased from M_n=12,000 (M_w/M_n=1.86) to M_n=86,000 (M_w/M_n=2.01) and, after cleavage of the grafted side chains, analysis of the graft units gave M_n=23,000 with M_w/M_n=1.26, in good agreement with the expected value of 21,000 [294].

Expanding on this idea, Hawker et al. copolymerized an alkene functionalized TEMPO derivative with propylene or 4-methylpentene using a cationic metallocene compound to prepare an α-olefin based macroinitiator for graft copolymerization (M_n=28,000, M_w/M_n=1.80 or M_n=6080, M_w/M_n=1.65) (Scheme 42) [295]. Heating the propylene-based copolymer to 123 °C in the presence of 200 equivalents of St increased its molecular weight to M_n=210,000 with M_w/M_n=

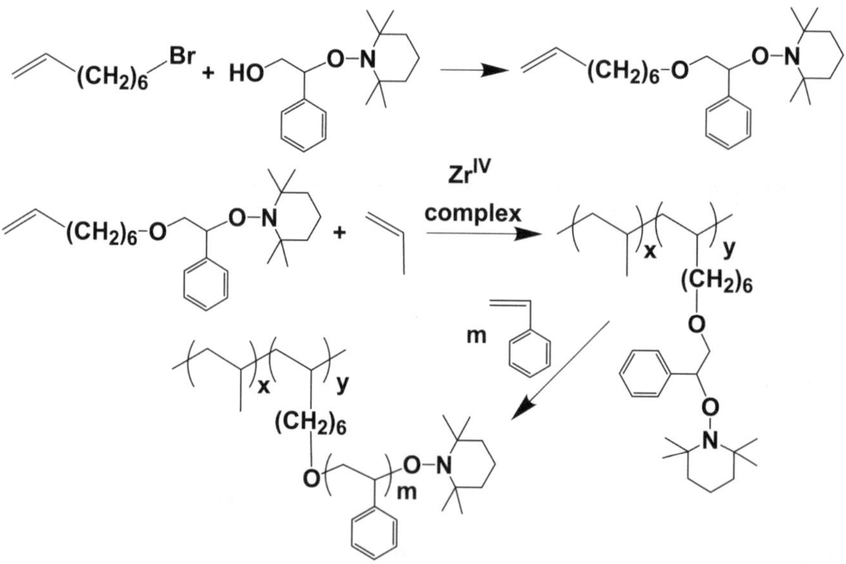

Scheme 42. Copolymerization of propylene and an alkene-functional TEMPO moiety followed by CRP of St to produce a graft copolymer [295]

2.0. The final molecular weights of the graft copolymers were dependent on the concentration of St added to the reaction. After cleavage, the molecular weight of the graft units ranged from $M_n=8000$ to $M_n=75,000$ with narrow molecular weight distributions ($M_w/M_n<1.43$) [295]. Chain extension of one of the graft copolymers with AcOSt resulted in an increase in molecular weight from $M_n=43,000$ ($M_w/M_n=1.70$) to $M_n=130,000$ ($M_w/M_n=1.95$). Chain extension with mixtures of St and BA (2:1) or St with MMA (3:1) were also successful. This indicated that the grafting procedure was well controlled and high chain end functionality was maintained throughout the reaction [295].

Shimada et al. later showed that commercially available polypropylene (pP) could be used as a macroinitiator for graft copolymerization once it had been functionalized with TEMPO moieties [296]. The pP was subjected to γ-irradiation, which formed peroxides in the chain that can cleave homolytically upon heating and generate radicals. These radicals were used for initiating the CRP of St in the presence of TEMPO. Using this method, however, a large amount of TEMPO must be added to control the polymerization, resulting in an excess of free TEMPO relative to the concentration of peroxide radicals in the system [296]. This excess was consumed by thermally initiated pSt chains, which resulted in a reliable probe of the molecular weight of the grafted units via analysis of the free chains, since the molecular weights of the free chains was controlled by the excess TEMPO. The experimental molecular weights of the grafted chains, as calculated from the weight of the grafted polymer divided by moles of peroxide units per gram of polymer, was in good agreement with the experimental molecular weights found for the free chains, suggesting that the grafted chains were formed in a controlled manner [296].

The grafting from technique can also provide a route to graft copolymers via ATRP. Matyjaszewski et al. used a poly [(vinyl chloride)-*co*-(vinyl chloroacetate)] copolymer, containing about 1% of the chloroacetate groups, to initiate the ATRP of St, MA, nBA, and MMA, resulting in graft copolymers with well-defined graft units(Scheme 43) [297].

^{1}H NMR analysis showed that the molecular weight of the macroinitiator increased substantially with each ATRP graft copolymerization ($M_n=47,400$ to

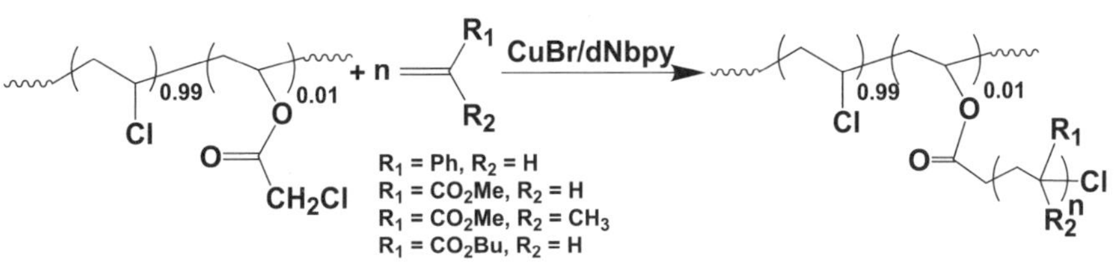

Scheme 43. Poly(vinyl chloride) graft copolymers via ATRP [297]

M_n=112,700–363,000) and that the mole percent of the grafted monomer ranged from 50–80% in the copolymer, because GPC analysis was not useful for determining the molecular weight of the copolymers [297]. A detailed analysis of the ATRP of nBA using ^{1}H NMR and FT-IR spectroscopies showed that with increasing reaction time, the mole percent of nBA in the copolymer increased and the T_g decreased. This work demonstrated that commercially available macroinitiators can be used to prepare graft copolymers via ATRP. In a similar way, grafting was performed from the defects in pVC by Percec et al. [298].

Sen et al. used either polyethylene or an ethylene styrene copolymer which was brominated using N-bromosuccinimide. The brominated polymers acted as initiators in the presence of a CuBr/PMDETA catalyst for the subsequent grafting of acrylic monomers. In similar studies, syndiotactic pSt-g-pMMA, syndiotactic pSt-g-pMA, and syndiotactic pSt-g-atactic pSt were synthesized by ATRP using brominated syndiotactic pSt as the initiator and the CuBr/PMDETA catalyst. Both the graft density and the MW of the graft segments were controlled by changing the bromine content of syndiotactic pSt and the amount of monomer used in the grafting reaction. The ATRP mechanism for grafting was supported by NMR analysis of the end groups. The thermal properties of the graft copolymers depended on both the graft density and the graft length [299].

Fónagy et al. used a commercially available polymer, poly(isobutylene-co-p-methylstyrene-co-p-bromomethylstyrene), as a macroinitiator for the ATRP of St to produce a graft copolymer [300]. After 23 h at 100 °C, with 1 equivalent of catalyst relative to the number of moles of initiator, monomer conversion was 81%; however, increasing the concentration of catalyst fivefold resulted in the same efficiency after only 11 h. They found no evidence of unreacted macroinitiator in the GPC analysis, indicating that nearly all of the chains had contained some grafted sites. Results from mechanical analysis showed that when the weight percentage of St was high (28 wt%), the graft copolymer exhibited no special properties; however, with only 6 wt% pSt, the polymer was elastomeric and could be reversibly stretched to 500% of its initial dimension [300].

Matyjaszewski et al. demonstrated that the grafting of St and MMA using a similar backbone is better controlled by applying the halogen exchange, as shown schematically in Fig. 39 [130, 301].

Mechanical and viscoelastic properties depend strongly on the morphologies of the samples, which are defined by the proportion of the hard and soft segments as well as by the uniformity of the hard pMMA grafts, as illustrated in Fig. 40 [301].

Pan et al. prepared a macroinitiator by chloromethylation of a commercially available AB block copolymer of poly(styrene-b-ethylene-co-propylene) (SEP) and used it as a macroinitiator for the ATRP of ethyl methacrylate (EMA) [302]. The kinetic plot showed little evidence of termination during the reaction and the molecular weight of the graft copolymer increased linearly with the monomer conversion, resulting in a final M_n=73,200 and M_w/M_n=1.22. The weight ra-

4 Other Chain Architectures   111

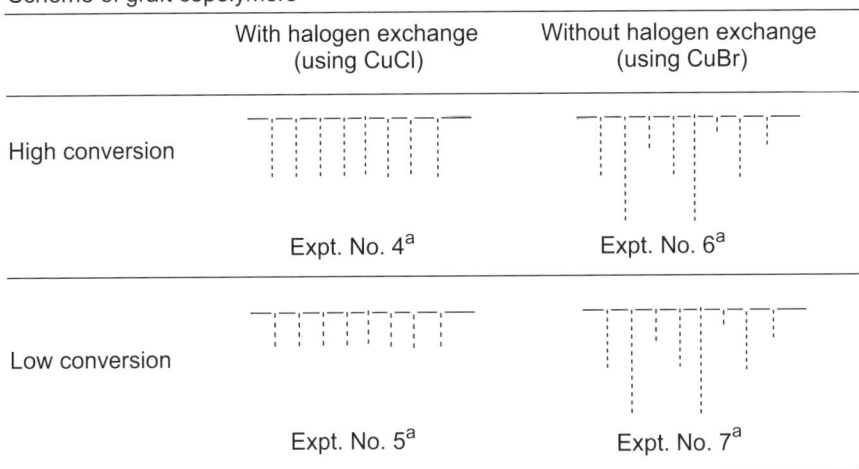

^{a}Experiment number refers to Figures 40 a and b

Fig. 39. Schemes for graft copolymers prepared from functionalized pIB

tio of pEMA to pSEP was 5.95:1 and DSC analysis showed the presence of three T_gs at -46, 78, and 103 °C, corresponding to the EP, EMA, and St segments, respectively [302].

A commercially available poly (ethylene-co-glycidyl methacrylate) (p(E-co-GMA)) polymer was transformed into an ATRP macroinitiator by Matyjaszewski et al. [303] by reacting the p(E-co-GMA) with either 2-bromoisobutyric acid or chloroacetic acid to prepare the functional backbone. The resulting polymer was used as a macroinitator for ATRP of St and MMA (Scheme 44). The consumption of both monomers increased with time, as did the weight percentages of the side chains in the copolymer. GPC analysis of the cleaved pSt side chains showed a linear increase of molecular weight with monomer conversion and $M_w/M_n<1.4$ [303]. ATRP has also been used in the grafting through process using hyperbranched polyethylene macromonomers with methacrylate functionality prepared by the living Pd-mediated process [304].

Inorganic macroinitiators can also be used for graft copolymerizations. Pendant vinyl functional pDMS was subjected to hydrosilation with 2-(4-chloromethylphenyl)ethyldimethylsilane to prepare a multifunctional ATRP macroinitiator by Matyjaszewski et al. [233, 234]. The ATRP of St was carried out using a pDMS macroinitiator with $M_n=6600$ and $M_w/M_n=1.76$ to yield a graft copolymer with $M_n=14,800$ and $M_w/M_n=2.10$. The increased polydispersity was attributed to the variation in the number of initiating sites on the pDMS backbone. ^{1}H NMR analysis showed that less than 5% of the total number of benzyl chloride moieties were left unreacted and that the weight ratio of pSt/pDMS was 1.18 [234].

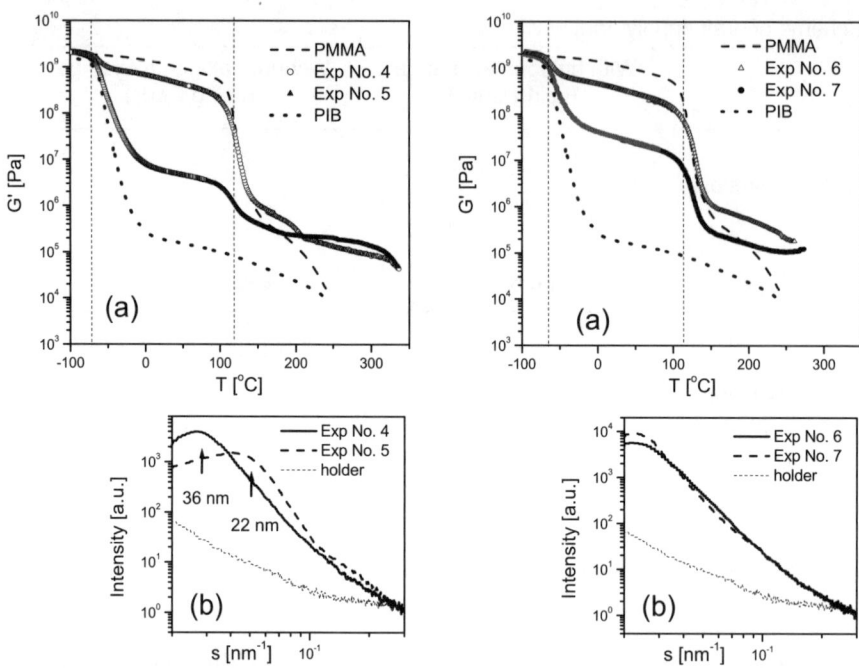

Fig. 40a,b. Viscoelastic properties of the pIB-g-pMMA with different compositions: The samples from Expt. No. 4, 5, 6 and 7 (description can be found in Fig. 39) in comparison with non-modified linear pIB and pMMA (a) and SAXS intensity distributions recorder for the same samples (b). Reprinted with permission from [301]. Copyright (2001) John Wiley & Sons, Inc.

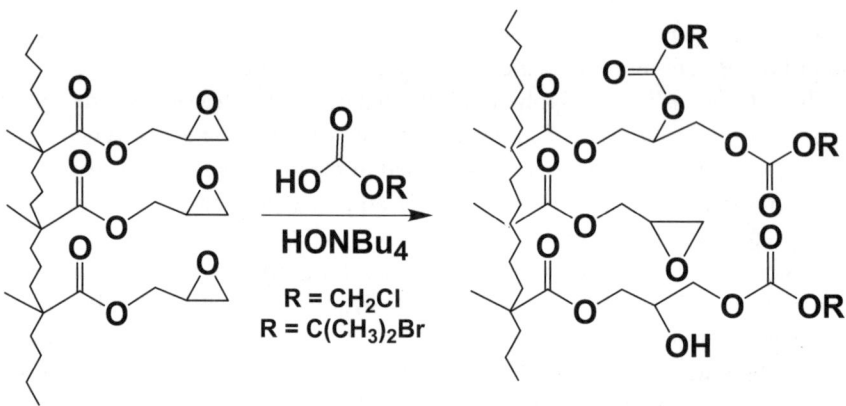

Scheme 44. Preparation of p(E-co-GMA) ATRP macroinitiator [303]

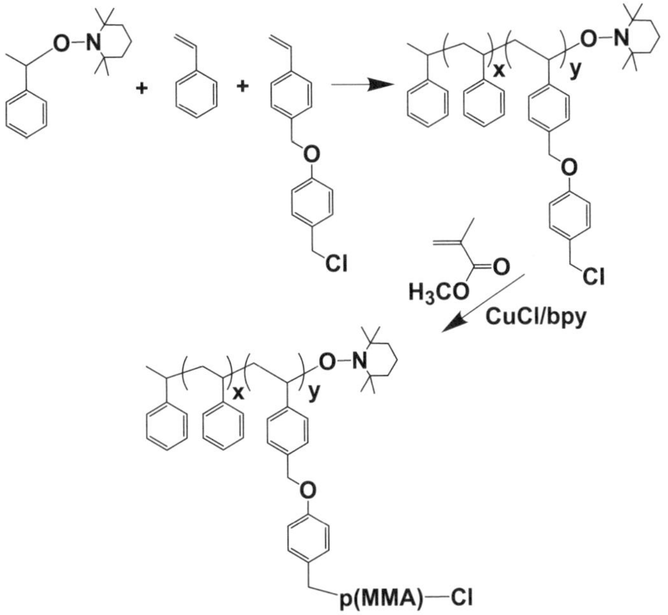

Scheme 45. Preparation of pSt-g-pMMA by tandem nitroxide-ATRP CRP methods [305]

Frechet et al. combined ATRP with nitroxide-mediated polymerizations to prepare graft and dendritic-type graft copolymers [305]. Copolymerizing St with p-(4′-chloromethylbenzyloxymethyl)styrene using a unimolecular TEMPO initiator resulted in a controlled polymerization producing a well-defined backbone. Subsequent ATRP of MMA, nBMA, or St using CuBr and various bpy ligands as the catalyst system provided the graft copolymers in relatively high yield with high molecular weights and fairly narrow molecular weight distributions (Scheme 45).

The use of alkyl-substituted bpy ligands in the formation of the catalyst complexes led to more well-defined polymers than when the unsubstituted bpy ligand was used [305]. The side chains of a pSt graft copolymer were cleaved with trimethylsilyl iodide and GPC analysis confirmed the controlled nature of the grafting reaction (M_n=15,000, M_w/M_n=1.29). By using a series of polymerizations, a dendrigraft copolymer was produced containing an initial pSt backbone, with pSt graft units containing pnBMA grafts. GPC analysis showed clean chain extension at each step, with little unreacted macroinitiator; however, the values obtained were lower than the theoretical predications. Higher molecular weights were obtained using a light scattering detector; these values were as expected. Thermal analysis showed the presence of two T_gs (30 °C (pnBMA) and 100 °C (pSt)), indicating that the dendrigraft polymer had been formed [305].

Liu et al. copolymerized TEMPO functional methacrylate- and St-based monomers with MMA and St, respectively, using ATRP conditions [306]. The resulting backbone copolymers were relatively well-defined (M_n=6200–8700 with M_w/M_n<1.65) and were used to initiate the polymerization of St at 120 °C for 24 h, apparently resulting in the graft copolymers [306]. No characterization data for the graft copolymers was provided.

The field of densely grafted copolymers has received considerable attention in recent years. The materials (also called bottle-brush copolymers) contain a grafted chain at each repeat unit of the polymer backbone. As a result, the macromolecules adopt a more elongated conformation. Examples of brush copolymers have been provided within the context of ATRP [307–309]. Synthesis of the macroinitiator was achieved through one of two approaches. One method used conventional radical polymerization of 2-(2-bromopropionyloxy)ethyl acrylate in the presence of CBr_4 to produce a macroinitiator with M_n=27,300, and a high polydispersity of M_w/M_n=2.3 (Scheme 46A) The alternative involved the ATRP of 2-trimethylsilyloxyethyl methacrylate followed by esterification of the protected alcohol with 2-bromopropionyl bromide. While synthetically more challenging, the latter method provided a macroinitiator of well-defined structure

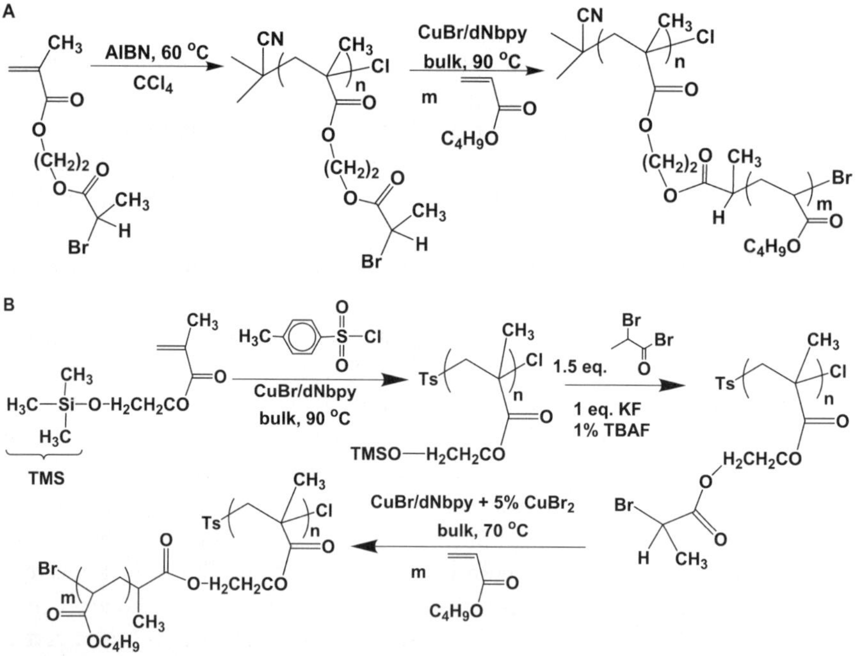

Scheme 46 A,B. Synthesis of densely grafted brush copolymers via: **A** a conventional radical polymerization followed by ATRP; **B** wholly by ATRP [307]

(M_n=55,500, M_w/M_n<1.3) leading to a brush synthesized entirely by a controlled process (Scheme 46B). From either macroinitiator, the ATRP of St and nBA was conducted, leading to the desired densely grafted structures. The grafting reactions were found to be very sensitive to reaction conditions; additional deactivator, high concentrations of monomer, and reduced temperatures were all necessary to produce the desired materials.

Since the aspect ratio and size of the macromolecules were so large, individual chains were observed by Atomic Force Microscopy (AFM) [307]. The brushes with pSt side chains form elongated structures on a mica surface with an average length of 100 nm, a width of 10 nm, and a height of 3 nm. pnBA absorbs well onto the mica surface and forms spectacular single molecule brushes in which the backbone and side chains can be visualized using tapping mode AFM [310]. Their shape depends on molecular structure and can be affected by applying pressure to these molecules [311]. They also exhibit very interesting properties in the bulk state [312].

Similarly, core-shell cylindrical brushes were prepared via block copolymerization [308, 313]. They consist of the soft pnBA cores and hard pSt shells [308]. The high resolution AFM micrographs of the block copolymer pBPEM-g-(pnBA-b-pSt) brushes shows a necklace morphology. The synthesis of well-defined brush block copolymers demonstrates the synthetic power of ATRP. It was used to create a well-defined backbone with a degree of polymerization of ~500, which was followed by a transesterification and the subsequent grafting of pnBA chains using ATRP. A final chain extension with St produced the block copolymers.

Doerffler and Patten have recently described a similar methodology for the formation of a less densely packed backbone where grafted polymers (macromolecules derived from only one monomer) were prepared strictly by ATRP [125]. The copolymerization of 4-acetoxymethyl- or 4-methoxymethylstyrene with styrene yielded a pendant functional macroinitiator with "latent initiation sites". Transformation of the ester or ether to benzyl bromide substituents provided the alkyl halide necessary for the grafting reactions. The increased polydispersities observed above 20% monomer conversion were attributed to internal coupling reactions between the grafted chains.

Kops et al. used cationic polymerization to produce a polyisobutylene-b-p(p-methoxystyrene) central block, which was then brominated to yield an ATRP macroinitiator (M_n=71,000, M_w/M_n=1.21) [314]. The extent of bromination was 44%, resulting in an initiating site at every fourth repeat unit. The ATRP of St was carried out using the CuBr/bpy catalyst system. The polymerization reached 14% monomer conversion after 4 h with a final M_n=244,000 and M_w/M_n=1.97. GPC using a light scattering detector showed the presence of a small fraction of high molecular weight polymer from termination by coupling. The ATRP of acetoxystyrene was more controlled (M_w/M_n=1.68) and no coupling products were observed [314].

There have been many examples in the literature of using a grafting from technique to obtain novel polymers and this method has proven to be robust and useful under a variety of reaction conditions. Table 10 summarizes the examples discussed in the above section.

Table 10. Examples of "grafting from" using CRP methods

Backbone	Grafts	Methods	Comments	Investigator
pSt	St	FRP/nitroxide	$M_n \uparrow$ 12,000→86,000	Hawker [372]
p(α-olefin)	St	Metallocene/nitroxide	$M_n \uparrow$, block copolymer side chains produced	Hawker et al. [295]
PP	St	Commercial/nitroxide	Contamination with thermally initiated pSt	Shimada et al. [296]
pVA	St	Commercial/ATRP	M_n=363,000, 80 mol% pSt	Matyjaszewski et al. [297]
pVA	MA	Commercial/ATRP	M_n=123,000, 50 mol% p(MA)	Matyjaszewski et al. [297]
pVA	nBA	Commercial/ATRP	M_n=227,000, 65 mol% p(nBA)	Matyjaszewski et al. [297]
pVA	MMA	Commercial/ATRP	M_n=162,000, 60 mol% p(MMA)	Matyjaszewski et al. [297]
p(IB-co-St)	St	Commercial/ATRP	81% grafting efficiency, 6–28 wt% pSt	Fónagy et al. [300]
p(SEP)	EMA	Commercial/ATRP	Wt ratio EMA:St=5.95:1, T_g= –46, 78, 103	Pan et al. [302]
p(ethylene)	St	Commercial/ATRP	69 wt% pSt, linear increase of graft MW with conversion	Matyjaszewski et al. [303]
p(ethylene)	MMA	Commercial/ATRP	80 wt% p(MMA), 2 T_gs, phase separation	Matyjaszewski et al. [303]
pDMS	St	Commercial/ATRP	$M_n \uparrow$ 6600→14,800, >95% grafting efficiency	Miller and Matyjaszewski [234]
pSt	MMA	Nitroxide/ATRP	M_n=119,000–141,000, M_w/M_n=1.19–1.79	Grubbs et al. [305]
pSt	nBMA	Nitroxide/ATRP	M_n=47,000–147,000, M_w/M_n=1.27–1.53	Grubbs et al. [305]
pSt	St	Nitroxide/ATRP	M_n=30,000–115,000, M_w/M_n=1.14–1.26	Grubbs et al. [305]
pMMA	St	ATRP/nitroxide	No characterization	Liu et al. [306]
pSt	St	ATRP/nitroxide	No characterization	Liu et al. [306]
pMMA	nBA	ATRP/ATRP	Graft unit at every other carbon	Matyjaszewski et al. [307]

4.1.2
Grafting Through

The grafting through technique can also be used to prepare graft copolymers. Hawker et al. demonstrated that copolymerizing preformed macromonomers (MMs) with St using a unimolecular TEMPO initiator produced controlled graft copolymers [315]. This approach allows incorporation of graft units derived from monomers that may not be polymerizable directly by CRP methods, but after suitable terminal functionalization can be incorporated as macromonomers. Well-defined (meth)acrylate terminated MMs composed of polycaprolactone (CL), poly(D, L)lactide, poly(ethylene glycol) and polyethylene, ranging in molecular weight from M_n=800 to M_n=8900, were copolymerized with St to produce graft copolymers. Specifically, when a pCL MM (M_n=2500, M_w/M_n=1.25) was copolymerized with St, the resulting graft copolymer had M_n=45,000 with M_w/M_n=1.35. The weight ratio of pCL determined from ^{1}H NMR was 28%, in good agreement with the feed ratio of 25 wt% [315]. Analysis of the copolymer backbone from a polymer where the graft units had been cleaved showed that the experimental molecular weight (M_n=37,500) was close to that predicted theoretically (M_n=43,000), illustrating that the reaction forming of the graft copolymer was well-controlled [315].

Similarly, Wang and Huang copolymerized St with a methacrylate-terminated pEO MM in the presence of AIBN/HTEMPO to prepare graft copolymers (Scheme 47) [316]. They found that the molecular weight of the copolymer increased with increasing reaction time and total monomer conversion, with M_w/M_n<1.35; however, the initial rate of polymerization decreased with increasing total monomer conversion, which was attributed to the increase in viscosity of the reaction medium at higher conversions. As the concentration of HTEMPO in the system increased, the molecular weight of the copolymer decreased and the molecular weight distribution narrowed, indicating that the HTEMPO controls the polymerization, as expected. The authors determined that, as the mo-

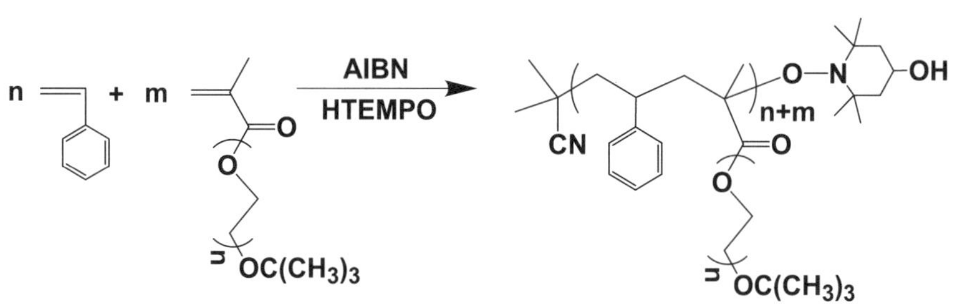

Scheme 47. Copolymerization of St with methacryloyl-terminated PEO to prepare well-defined graft copolymers [316]

lecular weight of the pEO MM increased, the reactivity ratio decreased, leading to less efficient incorporation [316]. Little polymer was produced when homopolymerization of the MMs was attempted, suggesting that incorporation into a polymer would also be feed dependent. An additional reason behind this behavior, although not addressed by the authors, is that homopolymerization of methacrylate MMs using the TEMPO-based systems would encounter the same problem that homopolymerization of low molecular weight methacrylate monomers does; the rate of decomposition of the end group competes with the activation process, leading to poorly controlled polymerizations.

Matyjaszewski et al. showed that ATRP can also be used to make well-defined MM graft units [317]. The ATRP of St was carried out using a vinyl chloroacetate initiator, which does not copolymerize with St under ATRP conditions, resulting in preparation of a well-defined MM (M_n=5800, 11,900, and 15,900, M_w/M_n= 1.12, 1.15, and 1.18). These MMs were copolymerized with N-vinyl pyrrolidinone (NVP) in a free radical polymerization to produce a graft copolymer (Scheme 48) [317]. The monomer feed ratio was varied from 10 to 50 wt% of NVP. Incorporation of the pSt MMs was both molecular weight and feed ratio dependent; as the molecular weight increased, the amount incorporated decreased and as the content of MM in the feed increased, the degree of incorporation decreased. There was little incorporation of the highest molecular weight MM. Both of these effects were attributed to an inaccessibility of the terminal vinyl bond, whether due to heterogeneity in the reaction medium or to steric hindrance that prevents the active site from approaching [317]. These graft copolymers swelled in water since the pNVP backbone is water-soluble while the pSt grafts are not, producing a hydrogel. Under equilibrium conditions the water content ranged from 74 to 97%, indicating that these copolymers may be useful as efficient absorbants [317].

Müller et al. used the MM technique to synthesize graft copolymers [318]. They copolymerized nBA with an ω-methacryloyl-pMMA MM both via a conventional radical process and an ATRP reaction. Although the MM was con-

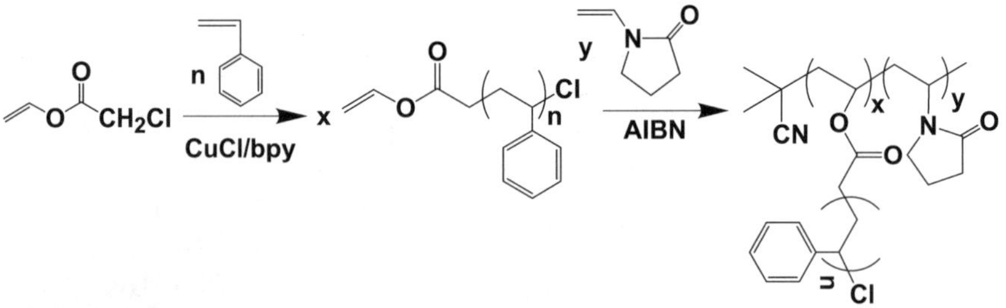

Scheme 48. Preparation of pNVP-g-pS by combining ATRP with a conventional radical process [317]

sumed faster than the nBA due to the higher reactivity of the methacryloyl chain end in the conventional radical process, the reactivity ratio for the MM was significantly lower than for the small molecule analog, MMA (r_1=1.6 vs 3.3) [318]. The rate of incorporation of the MM also decreased with increasing monomer conversion, as well as with increasing total monomers concentration, which was attributed to a decrease in the mobility of the MM in the reaction medium as the viscosity increased. In contrast, the reactivity ratio for the MM in the ATRP system was comparable to the reactivity ratio for MMA (r_1=2.2 vs 2.6) and the molecular weight of the graft copolymer increased linearly with increasing monomer conversion, indicating good control over the polymerization [318]. The enhanced reactivity of the MM in the ATRP system was attributed to the increase in the length of time between monomer additions to the polymer chain in ATRP vs a conventional process (~1 ms in conventional vs ~1 s in ATRP), which provides the MM with more time to reach the active center, lessening the effect of diffusion control over the reaction. In addition, the authors found that the reactivity ratios of the MM in ATRP were relatively unaffected by a change in the molecular weight of the MM or in the comonomer composition, at least up to M_n<11,000 [318, 319].

The controlled growth is even more important for pDMS MMs, which are less compatible with pMMA. Thus, ω-methacryloyl pDMS MMs in ATRP had reactivity ratios much closer to MMA than in a conventional process under similar conditions (r_{pDMS}=0.82 vs 0.34) [320]. The reactivity ratios depend on many factors, which include not only the polymerization mechanism but also the reaction temperature, as presented in Table 11 and Fig. 41 [139, 320].

Table 11. Copolymerization of MMA and pDMS macromonomer (PDMS-MA) [320]

Polymerization type	Initiator (I)[b]	MMA/PDMS-MA[c]/CDB/I molar ratio	Xylene wt%	Temp. °C	r_1[d]	$1/r_1$
RAFT	BPO	285/15/1/0.5	31	75	1.4±0.06	0.67
	AIBN	285/15/1/0.5	31	60	1.70±0.04	0.59
Conventional radical	AIBN	380/20/0/1	32	75	2.98±0.09	0.34
ATRP[a]	EBiB	285/15/0/1	31	75	1.37±0.10	0.73
	EBiB	285/15/0/1	31	90	1.25±0.01	0.81
	PDMS-Br	285/15/0/1	3	90	1.17±0.05	0.85

[a] Equimolar amount of CuCl [dnNbpy]$_2$ to initiator was used as the catalyst
[b] BPO: benzoyl peroxide, AIBN: 2,2′-azobisisobutyronitrile, EBiB: ethyl 2-bromoisobutyrate, pDMS-Br: pDMS macroinitiator containing 2-bromoisobutyrate end group (M_n=15,600, M_w/M_n=1.10, F=0.95)
[c] MMA/pDMS-MA composition in the feed: 95/5 (mol/mol), 45/55 (wt/wt). PDMS-MA: M_n=2370, M_w/M_n=1.25, F=1.0
[d] Reactivity ratio of MMA determined by Jaacks plot

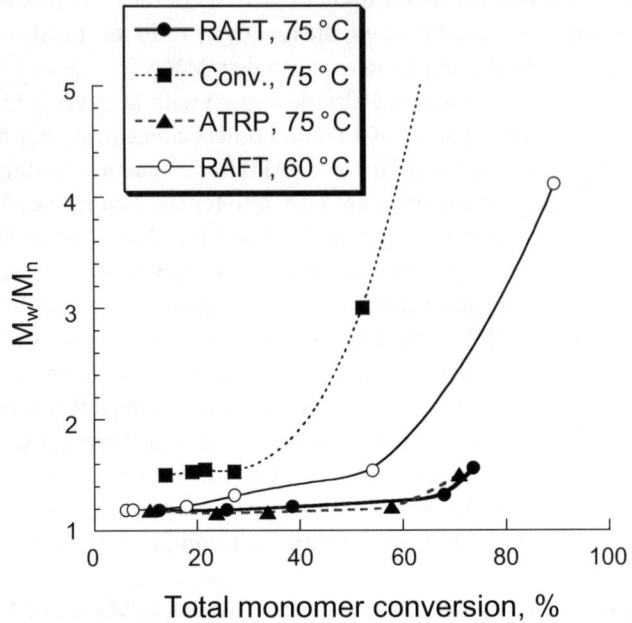

Fig. 41. Molecular weight distribution change for the copolymerization of MMA and pDMS macromonomer (pDMS-MA) (M_n=2370, F=1.0) in xylene solution (xylene=31 wt%). Conditions for RAFT: $[MMA]_0/[pDMS-MA]_0/[CDB]_0/[initiator]_0$=285/15/1/0.5, initiator/temperature: BPO/75°C (closed symbol), AIBN/60°C (open symbol). Conditions for the conventional AIBN polymerization: $[MMA]_0/[pDMS-MA]_0/[AIBN]_0$=380/20/1, temperature=75°C. Conditions for the ATRP: $[MMA]_0/[pDMS-MA]_0/[EBiB]_0/[CuCl]_0/[dnNbpy]_0$=285/15/1/1/2, temperature=75°C. Reprinted with permission from [320]. Copyright (2001) American Chemical Society.

A schematic representation of the distribution of grafts is shown in Fig. 42 and structure of macromonomers based on pDMS and pLA (see below) is in Fig. 43 [320, 321].

The most uniform graft distributions were obtained at higher temperatures and using pDMS macroinitiators, which compatibilize the growing chain and prevent phase segregation. Figure 44 illustrates the graft copolymers obtained in ATRP, RAFT, and a conventional RP and their mechanical properties. Graft copolymers with approximately the same MW (M_w=90,000) and the same overall composition (50 wt% of pDMS) but different branch distributions of pDMS grafts (M_w=2000) were prepared by three methods: ATRP (bottom Fig. 44), RAFT (top), and conventional RP (middle). ATRP provides a very uniform distribution of grafts and RAFT yields a copolymer with a gradient structure, whereas a conventional RP leads to copolymers with broad MWD and differences not only along the chain length but among different chains, since it is not a

4 Other Chain Architectures 121

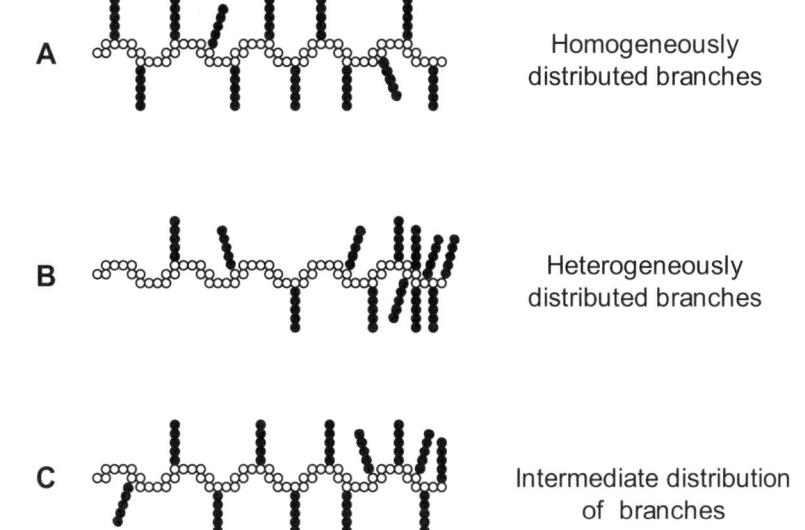

Fig. 42. Schematic representation of the types of graft copolymers resulting from the copolymerization of small molecules with pDMS macromonomers

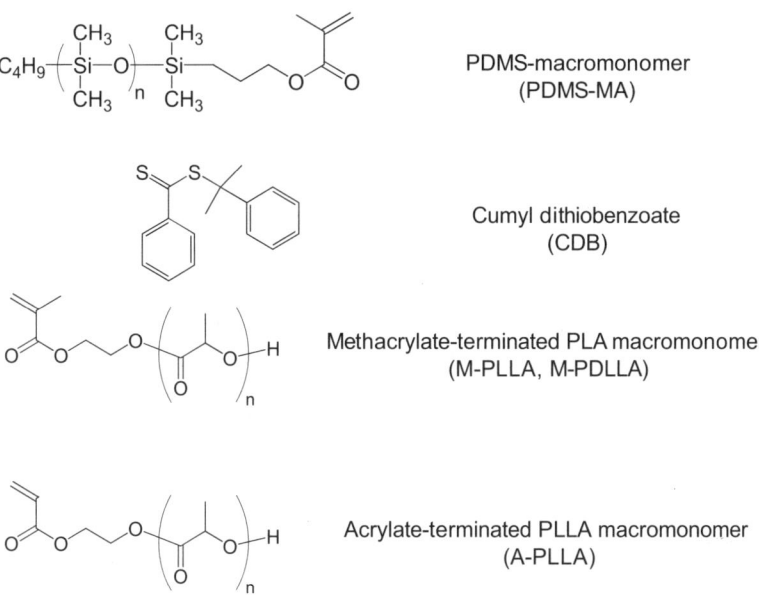

Fig. 43. Structures of pDMS and pLLA macromonomers used to prepare graft copolymers

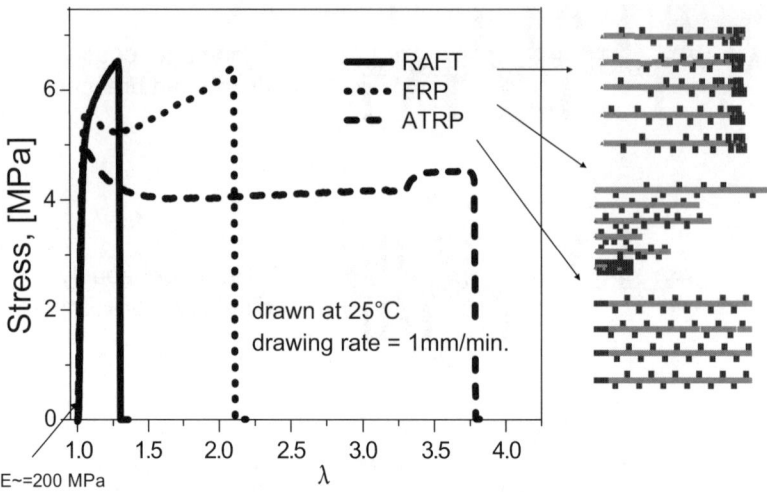

Fig. 44. Right: copolymers of pDMS methacrylate macromonomers with different branch distributions prepared by different methods: RAFT (top, solid line), FRP (middle, dotted line), ATRP (bottom, broken line). Left: impact of branch distribution on mechanical properties (stress vs draw ratio) [320]

living method. Depending on the method of preparation, graft copolymers have dramatically different mechanical properties, which may also lead to variable and adjustable surface properties [139]. As shown in Fig. 44, copolymers with different branch distributions have very different tensile elongations at the break point. Cold drawing of the graft copolymer prepared by ATRP at 25 °C (broken line) leads to 280% tensile elongation, that prepared by RAFT (solid line) a 30% tensile elongation, whereas the copolymer prepared by the conventional radical process breaks at 115% elongation. This demonstrates the dramatically different properties of the copolymers with the same overall composition.

Poly (lactic acid) (LA) macromonomers behave in a different way [321]. The intrinsic reactivity of methacryloyl terminated pLA is higher than that of MMA, whereas the acryloyl terminated pLA derivative is less reactive. They follow the reactivities of the corresponding hydroxyethyl derivatives, as shown in Table 12.

pLA is miscible and compatible with pMMA. However, the frequency of addition of the MMs is still low, which affects the polydispersity of the graft copolymers and the reactivity ratios, as shown in Table 12 and Figs. 45 and 46. Well-defined graft copolymers with low polydispersities have been prepared, but the graft distribution was unsymmetrical and showed a gradient for both acryloyl and methacryloyl pLA derivatives. Apparently, simultaneous copolymerization of both of them with MMA provides a uniform structure, as shown schematically in Fig. 47 [321].

Table 12. Reactivity ratios for the copolymerization of MMA (M1) with comonomers (M2) [321]

M_2	Polymerization system	Solvent[h]	$[M_1]_0$, mol/l	Temp. °C	r_1[i]	$1/r_1$
M-PLLA[a]	ATRP	DPE/xylene	2	90	0.57±0.02	1.75
M-PDLLA[b]	ATRP	DPE/xylene	2	90	0.68±0.17	1.47
HEMA[c]	ATRP	xylene	4	90	0.67±0.02	1.49
HEMA[c]	Conventional (AIBN)	(bulk)	9	80	0.75	1.33
EMA[d]	Conventional (AIBN)	dioxane	1	60	0.85±0.01	1.18
M-PLLA[e]	Conventional (AIBN)	dioxane	1	60	1.01±0.17	0.99
M-PLLA[a]	Conventional (BPO)	DPE/xylene	2	90	1.09±0.05	0.91
A-PLLA[f]	ATRP	DPE/xylene	2	90	1.63±0.10	0.61
HEA[g]	ATRP	xylene	2	90	1.57±0.07	0.64

[a]Methacrylate-terminated poly(L-lactic acid) macromonomer, M_n=2800
[b]Methacrylate-terminated poly(D,L-lactic acid) macromonomer, M_n=3350
[c]2-Hydroxyethyl methacrylate
[d]2-Acetoxyethyl methacrylate
[e]Methacrylate-terminated poly(L-lactic acid) macromonomer, M_n=4500
[f]Acrylate-terminated poly(L-lactic acid) macromonomer, M_n=2690
[g]2-Hydroxyethyl acrylate
[h]DPE: diphenyl ether
[i]Determined by Jaacks method

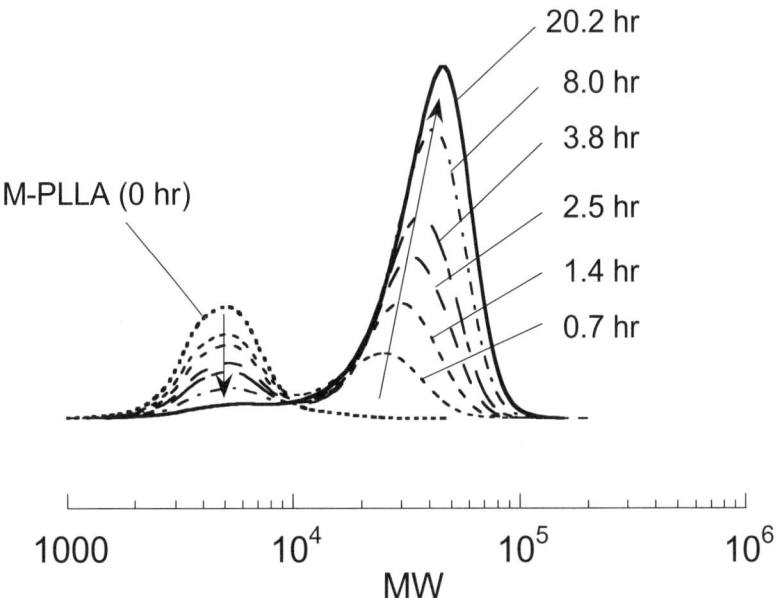

Fig. 45. GPC traces for the ATRP of MMA and M-pLLA (M_n=2800, M_w/M_n=1.16, F=1.0). Conditions: $[MMA]_0/[M\text{-}pLLA]_0/[EBiB]_0/[CuCl]_0/[dnNbpy]_0$=289.5/10.5/1/1/2, xylene= 21wt%, diphenyl ether=36wt%, 90°C, under N_2. Reprinted with permission from [321]. Copyright (2001) American Chemical Society.

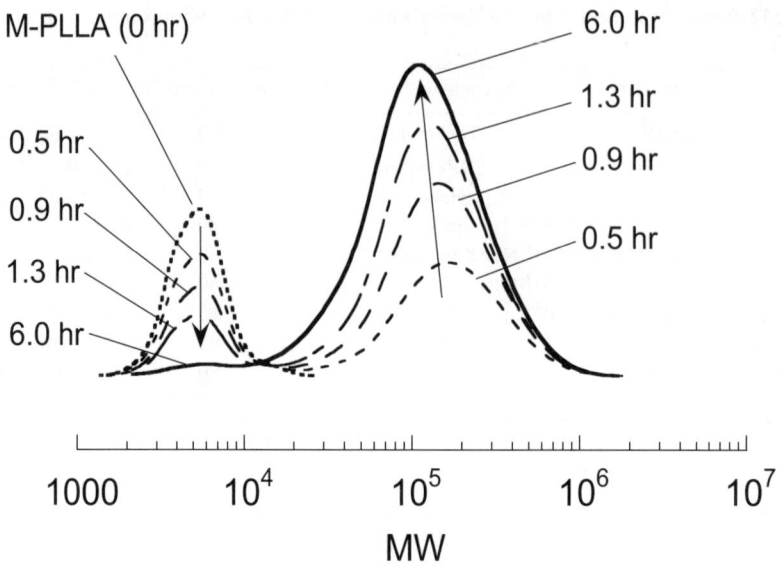

Fig. 46. GPC traces for the conventional radical polymerization of MMA and M-pLLA (M_n=2800, M_w/M_n=1.16, F=1.0). Conditions: $[MMA]_0/[M\text{-}pLLA]_0/[BPO]_0$=289.5/10.5/1, xylene=21wt%, diphenyl ether=36wt%, 90°C, under N_2. Reprinted with permission from [321]. Copyright (2001) American Chemical Society.

In a similar way, *n*-butyl acrylate was copolymerized by ATRP with methacrylate macromonomers containing highly branched polyethylene prepared by Pd-catalyzed living ethylene polymerization. The observed reactivity ratios depend on the molecular weight and concentration of the macromonomer. The resulting graft copolymers showed microphase separation by AFM [304].

Fukuda et al. took the MM method a step further and used ATRP to homopolymerize an ω-methacryloyl-poly (isobutyl vinyl ether) (IBVE) MM [309]. After preparing the pIBVE-MM via a cationic process, they used ethyl 2-bromoisobutyrate as the initiator and CuCl/4,4′-diheptyl-2,2′-bipyridine (dHbpy) as the catalyst for the ATRP reaction. The MMs were well-defined with M_n=1600 to 5300 and M_w/M_n=1.04 to 1.07. The ATRP of a pIBVE-MM (M_n=1600, M_w/M_n=1.07) at a 30:1 ratio of monomer:initiator was carried out in 50% diphenylether at 50 °C and proceeded with little termination, reaching 90% MM conversion after 5 h [309]. The rate of the polymerization was constant up until about 70% MM conversion, then decreased slightly, which was attributed to the increased viscosity of the reaction medium, as in the above MM examples. The molecular weights of the graft copolymers increased linearly with conversion and the molecular weight distributions remained narrow (M_w/M_n<1.2). Higher molecular weight graft copolymers (DP≥100) could be prepared once the catalyst concentration was increased, with the polymerizations reaching >85% MM conversion after 24

4 Other Chain Architectures

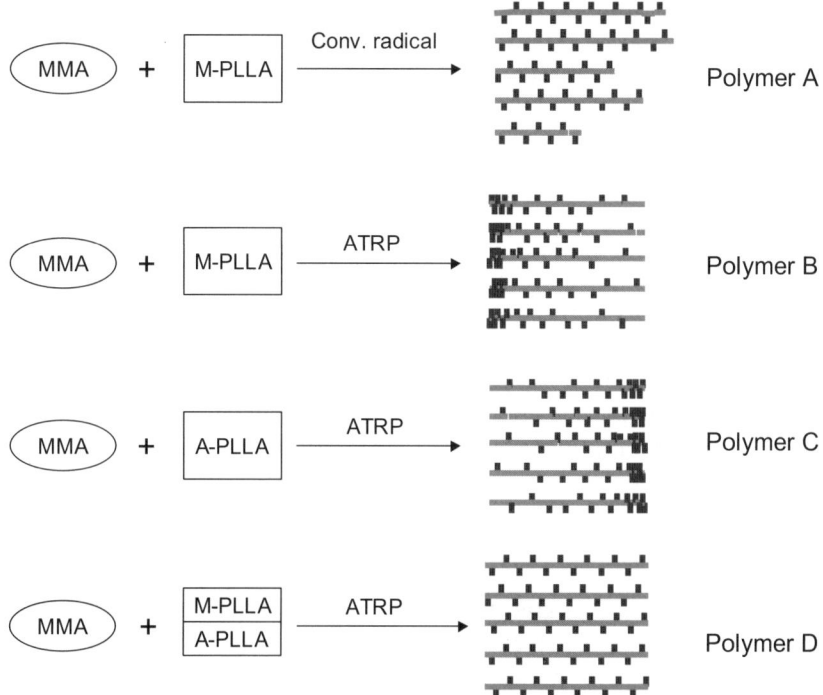

Fig. 47. Illustration of the growth of polymer chains using the MM method and various radical polymerization techniques. Reprinted with permission from [320]. Copyright (2001) American Chemical Society.

Table 13. Graft copolymers prepared using "grafting through" and CRP methods

MM	Comonomer	Comments	Investigator
PCL;pLA;pEG	St	pCL: M_n=2500; pSt-g-pCL: M_n=45,000; 28% pCL	Hawker et al. [315] Matyjaszewski [321]
pEO	St	MW of pEO ↑, % incorporation ↓	Wang and Huang [316]
pSt	NVP	MW of pSt ↑, % incorporation ↓, hydrogels, H$_2$O (eq)=74–97%	Matyjaszewski et al. [317]
pMMA	nBA	r_1 (pMMA)=2.2 in ATRP, close to MMA monomer	Müller et al. [318]
pIBVE	–	M_n=1600, DP=30, conv.=90%, t=5h; DP=100, conv.=85%, t=24 h; DP=200, conv.=87%, t=110h	Fukuda et al. [309]
pDMS	MMA	ATRP, RAFT	Matyjaszewski [139, 320]

h. At a DP=200, the polymerization took 110 h, but the MM conversion was 87% with M_n=99,200 and M_w/M_n=1.19. The ATRP copolymerization of the higher molecular weight MMs was also controlled; however, for the highest molecular weight MM (M_n=46,000) at DP=30, the polymerization only reached 26% MM conversion after 22 h, indicating that the rate of the polymerization was significantly affected by the molecular weight of the MM [309].

There have been a significant number of literature reports that have utilized the grafting through technique to prepare copolymers [322]. Unfortunately, this technique has had some limitations. The maximum MM conversion is strongly affected by the ability of the MM to diffuse toward the active center. Increasing the time between monomers additions, which can be accomplished by using a CRP process, has lessened this effect, but homopolymerizations of MMs is still challenging due to viscosity effects. Table 13, in addition to Table 12 and Fig. 47, summarize the attempts made at using CRP for grafting through reactions.

4.1.3
Grafting Onto

Hawker et al. explored the grafting onto technique to examine the steric effects associated with coupling polyether dendrons to a functional backbone [323]. St was copolymerized with N-oxysuccinimide 4-vinylbenzoate, which contained an active ester moiety, using a phenylethyl-TEMPO derivative as a unimolecular initiator, resulting in formation of copolymers with 10–40% of the active ester moiety in the backbone and molecular weights ranging from M_n=43,000 to 65,000. Graft copolymers were prepared by dissolving the backbone and a slight excess of the amino-functional dendrons in chloroform, then heating the mixture to reflux for 12 h (Scheme 49). GPC analysis of the graft copolymer after precipitation, which removed the excess dendron, showed that the molecular weight increased significantly from that of the starting backbone, as well as from that of the dendron, with no evidence of residual unreacted species. Although GPC analysis gave qualitative information regarding the molecular weights, ^{1}H NMR was used to determine the weight percent of the dendrons as well as the number of remaining ester functionalities in the backbone and from there the actual molecular weights of the graft copolymers were calculated. These values were significantly higher than those determined against linear pSt standards and the deviation was larger for copolymers incorporating both larger dendrimers and at higher weight percentages, as would be expected for this type of three-dimensional shape [323]. After accurately determining the molecular weights, the authors investigated the steric effects associated with the grafting onto approach. They found that even at high loading capacities (>90 wt% dendrimer), the backbones containing 10–20% of the active esters were completely functionalized. However, once the content of active esters increased to 30 and then to

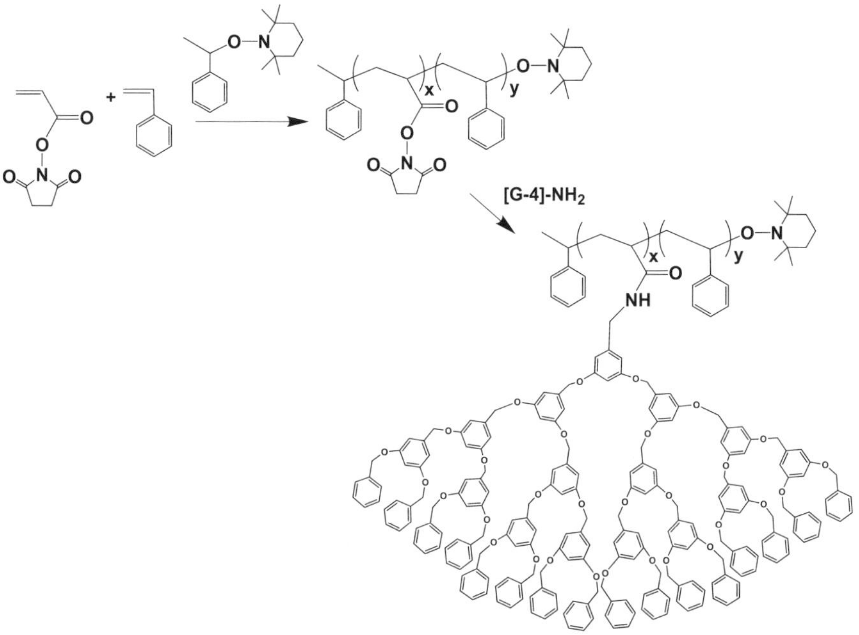

Scheme 49. Synthesis of graft copolymers containing dendrimers via a "grafting to" approach [323]

40%, there was significant steric hindrance, preventing the higher generation dendrons (>G-3) from reacting with the backbone to form the graft copolymer [323].

4.1.4
Grafting from Surfaces

4.1.4.1
Silicon and Gold

In a twist on the grafting from technique, investigators have begun to use CRP methods to grow polymers off functionalized surfaces [181]. Both nitroxide mediated polymerization and ATRP have been used for this purpose. In 1998, Wirth et al. used ATRP to grow acrylamide off the surface of silica capillaries functionalized with benzyl chloride moieties to prepare coatings for capillary electrophoresis [324]. After covalently attaching the benzyl chloride initiating sites to the silica surface, the ATRP of acrylamide was carried out using the CuCl/bpy catalyst in 50% DMF solution at 130 °C over 40 h, with and without cross-linker present. AFM analysis of the films prepared using the surface bound initiator showed that they were much more uniform than films prepared by solution polymerization of the monomers and simply adsorbed onto the surface. The integrity of the capillary coatings produced were then investigated. Both the linear and crosslinked films were able to successfully separate three distinct proteins at pH 4.5 and the reproducibility of the results had less than 1% error after 20 runs. After 150 runs (2 weeks) the elution times of the proteins were slightly shorter, the crosslinked film producing results within the standard deviation [324]. The difference between the integrity of the linear and crosslinked films was attributed to slow hydrolysis of the linear film, resulting in an increase in the electro-osmobility, which is indicative of irreproducible results [324]. Later work investigated the distribution of chain lengths resulting from this surface growth and found narrow molecular weight distributions, M_w/M_n <1.5, suggesting that the grafting from reactions were indeed controlled [325].

Fukuda et al. showed that Langmuir-Blodgett techniques can be used to form a monolayer of 2-(4-chlorosulfonylphenyl)ethyl trimethoxysilane on a functionalized silicon surface by dragging the surface through the monolayer (Scheme 50) [326]. This treated surface was then used to initiate the ATRP of MMA using the CuBr/dHbpy catalyst system in the presence of an untethered initiator, p-toluenesulfonyl chloride (TsCl). There was a linear increase in the surface thickness as a function of the reaction time and with the molecular weight

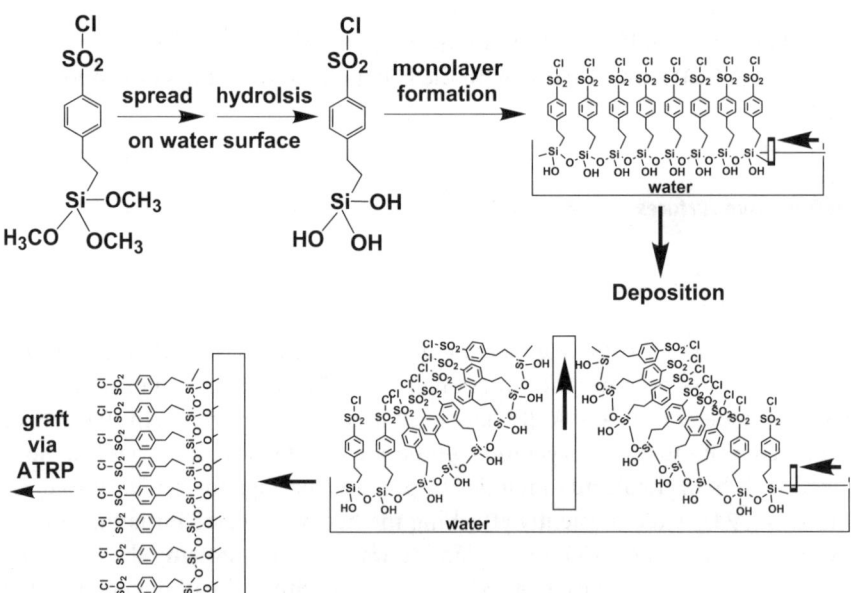

Scheme 50. Illustration of immobilization of an ATRP initiator onto a surface using Langmuir-Blodgett techniques [326]

of the free chains generated by the added TsCl, suggesting that the growth from the surface was controlled [326].

In 1999, Hawker et al. prepared TEMPO moieties containing reactive groups that could be used to tether the initiator to silicon surfaces (wafers or gel particles) [327]. One of the major difficulties associated with growing the polymers off the surfaces, which Wirth et al. [324]. had not addressed but that Fukuda et al. [326] had considered, is the extremely low concentration of initiating sites. This leads to a low concentration of persistent radicals (i.e., free nitroxide or Mt^{n+1} for ATRP) in the contacting solution and leads to an uncontrolled polymerization. Hawker et al. added a small amount of the 1-phenylethyl-TEMPO to the system and were able to control the polymer growth from the surface [327]. The free polymer chains were separated from those attached to the surface by washing the surface with an appropriate solvent. To demonstrate that the polymer does indeed grow off the surface, St was polymerized and the thickness of the surface was monitored as a function of the molecular weight of the pSt chains in solution (which was assumed to be the same as that of the free chains). There was a linear relationship between the M_n of the pSt vs the increase in the layer thickness, suggesting that the process is controlled. Furthermore, the thickness also increased with increasing monomer conversion, as would also be expected for a process where all the chains grow simultaneously [327]. To obtain even more information regarding the extent of control, initiator moieties containing a benzyl ether linkage were used and after the polymerization of St, the chains were cleaved and their molecular weight characteristics determined. The molecular weight of the cleaved chains using a 500:1 ratio of monomer: initiating sites was $M_n=51,000$ ($M_w/M_n=1.14$), in good agreement with the molecular weight of the free chains that were isolated ($M_n=48,000$, $M_w/M_n=1.20$). The lower molecular weight distribution in the grafted chains, as opposed to the free chains, was attributed to the absence of chains resulting from thermal initiation. To confirm that the tethered polymer chains still contained an active TEMPO chain end, the chains were extended with a 1:1 mixture of MMA and St at a ratio of 250:1 of the monomer: initiating sites, which resulted in an increase in thickness of about 26 nm for all the surfaces investigated [327]. IR analysis confirmed the presence of the MMA units on the surface. Random copolymers of St with HEMA were also prepared, which altered the hydrophilic nature of the surface. ATRP from the surface was also attempted. A bromoisobutyrate functional group tethered to the surface was successfully used for the ATRP of MMA in conjunction with $NiBr_2(PPh_3)_2$ as the catalyst and in the presence of ethyl 2-bromoisobutyrate as the free initiator. As had been seen in the nitroxide-mediated polymerization of St, the surface thickness increased with the increasing molecular weight of the free chains, indicating that the process was controlled [327].

Matyjaszewski et al. used ATRP to grow first well defined block copolymer chains from silicon surfaces by CRP [328]. They used a process similar to that of

Hawker et al. [327] where 2-bromoisobutyryl functional initiators were first coupled via long alkyl linkers to silicon wafers [328]. The presence of the halogens on the surface was confirmed using X-ray photoelectron spectroscopy (XPS). The ATRP of St was carried out in the absence of free initiator, but with 5 mol% of $CuBr_2$/dNbpy relative to the CuBr/dNbpy added to the solution to control the polymerization. There was a linear correlation between the film thickness on the surface and the expected molecular weight of the grafted chains (based on the results of the polymerization performed under the same conditions but with free, instead of surface bound initiator). XPS spectroscopy confirmed a decrease in the intensity of the Si and O signals originating from the surface, suggesting that the surface had become covered with the organic pSt layer. There was also a 50% reduction in the signal for the bromine, which could result either from loss of end-group functionality or from chains that are not fully extended and therefore the chain ends are not readily detectable by the XPS measurement [328]. Additional Cu(II) was required in order to provide a sufficient amount of deactivator present to allow a controlled polymerization. Reactions performed without added Cu(II) showed that the thickness of the surface film was approximately five times larger than when it was present, and film thickness did not increase with increasing reaction time, suggesting the growth from the surface had not been controlled. The ATRP of MA from the surface was also carried out, resulting again in a linear increase of the film thickness with reaction time. To verify that the chain ends remained active and that XPS was an ineffective measure of the remaining bromine content, the surface coated with tethered pSt was chain extended with MA. The surface film thickness increased with reaction time, with a final thickness greater than 100 nm [328]. To demonstrate that the surface characteristics could be altered, block copolymers were prepared (Scheme 51). The pSt-coated surface was chain extended with tBA, resulting in an increase in the thickness from 26 nm to 37 nm after 4.5 h. The presence of the tBA was confirmed through internal reflectance IR spectroscopy. The *tert*-butyl esters were hydrolyzed to form the corresponding acid groups, resulting in a surface that had a water contact angle of 18°, as compared to the 86° for the original pSt surface, suggesting that the preparation of tethered block copolymers may be a successful approach for tuning surface properties [328]. To demonstrate further that the surface properties can be significantly altered by simple changes in the chemistry, a fluoroacrylate monomer was polymerized from the surface, which resulted in a large increase in the water contact angle, to 119°.

Zhao and Brittain also used the grafting from technique to grow polymers off silicon surfaces [329]. In their example, however, they first prepared tethered pSt cationically, which was terminated with a chlorine chain end, then used it as the initiator for the ATRP of MMA using the CuBr/PMDETA catalyst system. Use of a chlorine-terminated pSt chain in conjunction with a CuBr catalyst, however, produces a mismatch between the rates of cross-propagation and the rate of po-

4 Other Chain Architectures

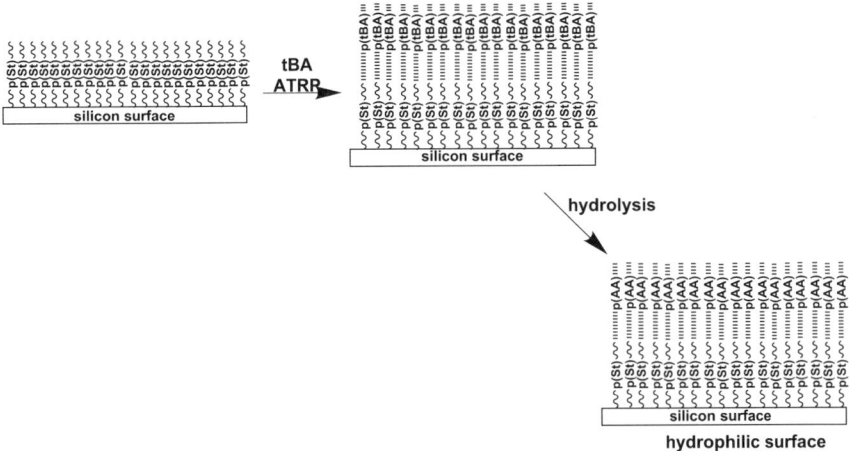

Scheme 51. Illustration of growing pSt-ptBA block copolymers from a silicon surface using ATRP methods and conversion to pSt-pAA [328]

lymerization in the ATRP polymerization of MMA, and may lead to poorly defined polymers and inhomogeneous surfaces [176]. They followed the St polymerizations by reflectance IR spectroscopy and found that in the initial polymerization step, not all the initiator was consumed, but upon the chain extension with more St, the absorbance associated with the initiator species decreased, resulting in an increase in the film thickness from 34 nm to 42 nm [329]. The ATRP of MMA initiated by the surface tethered pSt chains (26 nm thickness) led to an increase in the film thickness to 35 nm. When the surfaces were exposed to different solvents, which were selected either for the pSt or the pMMA blocks, the water contact angles changed accordingly. This was confirmed through the use of XPS and AFM analysis, which showed that the nature of the surface changed when it was exposed to different solvents [330]. Again, this demonstrates that the surface properties can be tuned for specific applications.

As first discussed in the block copolymer section, ATRP can be approached from either side of the equilibrium. Brittain et al. took this approach to tether polymer chains to a polymer surface [331]. After functionalizing a silicon surface with diazo moieties, the rATRP of St was carried out using the $CuBr_2$/PMDETA catalyst system in an anisole solution at 90 °C for 24 h. This resulted in a film 25 nm thick. In the absence of the $CuBr_2$, the film thickness was 70 nm, supporting the idea that the $CuBr_2$ acts as a mediator for the polymerization and controls the growth of the polymer chains from the surface [331]. Using a conventional system for further ATRP chain extension with MMA (CuBr/PMDETA, anisole solution, 90 °C, 24 h), they produced an increased film thickness (31.5 nm) and a decrease in the water contact angle (74° vs 92°) suggesting that the pMMA was indeed attached to the pSt chains [331]. However, in contrast to Hawker et al. [327], Matyjaszewski et al. [328], and Tsujii et al. [326], who all insisted that control could not be achieved in the absence of either sacrificial initiator or excess deactivator, neither the initial report using sequential cationic/ATRP methods nor the report using rATRP followed by the conventional process used either of these additives to control the polymerization.

Another type of surface that has been explored as a medium for tethered surface polymerization is gold. Hawker et al. and Baker et al. showed that self-assembled monolayers (SAMs) of halogenated alkyl thiols on gold were capable of initiating the ATRP of several different types of monomers [332, 333]. Hawker et al. used a 2-bromoisobutyrate functionalized surface in the presence of a small amount of untethered initiator and $FeBr_2(PPh_3)_3$ as the catalyst for the ATRP of MMA at 60 °C. They found that the brush thickness on the surface correlated well with the molecular weight of the polymer chains grown in solution and that, based on the molecular weight and the contour length of the tethered brushes, the grafting density was about one chain per ten potential initiating sites on the monolayer, indicating a high density of brushes on the surface [332]. After the successful ATRP of MMA, brushes were also prepared from *tert*-butyl, isobornyl, hydroxyethyl, and (dimethylamino)ethyl methacrylates. The measured water contact angles were a function of the hydrophilicity of the surface, as had been found for other systems [327, 328]. The hydrophobic surfaces were subsequently analyzed for their resistance to etching, which was dependent on the type of etching agent used, indicating that this surface property can be tuned to specific applications [332].

Baker et al. condensed 2-bromopropionyl bromide to surfaces containing 11-mercaptoundecanol to prepare SAMs capable of initiating ATRP (Scheme 52)

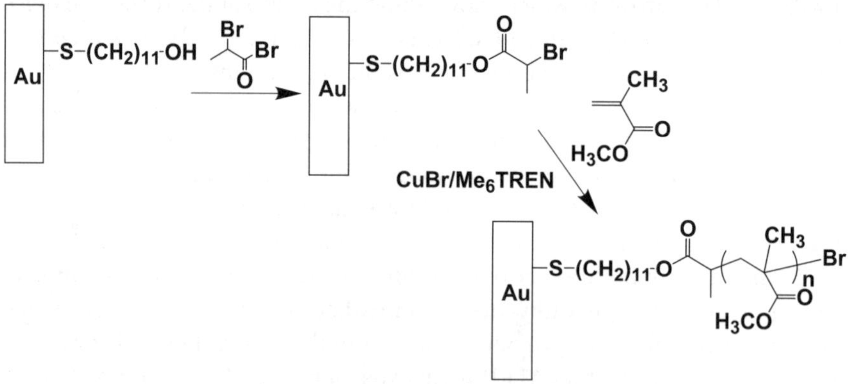

Scheme 52. Functionalization of a gold surface with a 2-bromopropionyl initiator capable of participating in the ATRP of MMA [333]

[333]. They utilized a CuBr/Me$_6$TREN catalyst system for the ATRP of MMA, which allowed them to achieve polymerization at lower temperatures and avoid the decomposition of the SAMs that is common at higher temperatures. After carrying out the ATRP reaction at 25 °C for 12 h, the film thickness increased from 12 Å to 370±5 Å [333]. GPC analysis of the brushes cleaved using an I$_2$ treatment gave M$_n$=44,500 with M$_w$/M$_n$=1.30, showing the polymerization had been well-controlled.

The synthesis of dispersed silicate nanocomposites was achieved through the CRP of St within layers of alkoxyamine-loaded montmorillonite (Scheme 53) [334]. Organic/inorganic nanocomposites of intercalated and delaminated silicates containing polymers is an area of continued interest owing to the enhanced thermal and dimensional stability these materials possess. While a variety of techniques have been developed for the synthesis of polymer layered silicate nanocomposites, the preparation of dispersed silicate nanocomposites has not been as extensively developed. However, using CRP, dispersions of clay particles within a matrix of well-defined pSt were formed. An alkoxyamine with a pendant quaternary ammonium group was synthesized and after exchange with sodium cations from the pristine clay, SFRP initiators could be loaded between the silicate layers. X-ray diffraction (XRD) patterns provided evidence for the loading and intercalation of the alkoxyamines between the silicate layers where the interlayer distance increased from d=1.26 nm (Na$^+$ spacing) to d=2.35 nm (alkoxyamine spacing). The subsequent addition and polymerization of St at 125 °C in the presence of the functional montmorillonite resulted in complete delamination of the silicate layers. XRD provided further evidence for the delamination and dispersal of silicates by the disappearance of the diffraction pattern after the SFRP of St. Ion exchange and extraction of pSt (M$_n$=21,500;

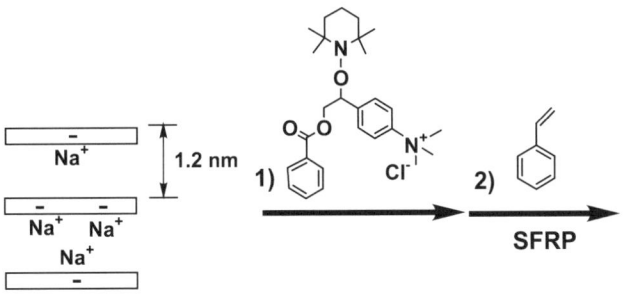

Layered Silicate **Dispersed Silicate Nanocomposite**

Scheme 53. Synthesis of dispersed silicate nanocomposites from SFRP of St within montmorillonite [334]

$M_w/M_n=1.3$) from the dispersed nanocomposite revealed that well-defined polymers were made, with M_n values in close agreement with theoretical predictions. A similar approach applied ATRP [335].

4.1.4.2
Grafting from Particles

Hybrid materials composed of inorganic nanoparticles and organic surface groups possess interesting optical, magnetic, and blending properties. These hybrids containing nanoparticles have been prepared by several synthetic routes: trapping colloids within a crosslinked matrices, "grafting to" particles with functional molecules/polymers, or "grafting from" particles using a living, or controlled polymerization process. CRP functionality has also been introduced to colloidal materials by the attachment of ATRP initiating groups to the particle surface. The properties of the hybrid nanoparticles prepared from this method can be tuned by varying the particle size of the colloidal initiator, changing the composition of the particle core, or by tethering (co)polymers with novel compositions/functionalities. An interesting feature of hybrid nanoparticle ultrathin films has been the formation of ordered two-dimensional arrays of particles, with a spacing dependent on the radius of gyration of the tethered (co)polymer. The general method for the synthesis of silsesquioxane-based hybrid nanoparticles and ATRP is presented in Scheme 54.

Von Werne and Patten, and Matyjaszewski et al., respectively, prepared siloxane-based nanoparticles via the base-catalyzed hydrolysis and condensation of tetralkoxysilanes (i.e., the Stöber process) [336, 337] or using a microemulsion

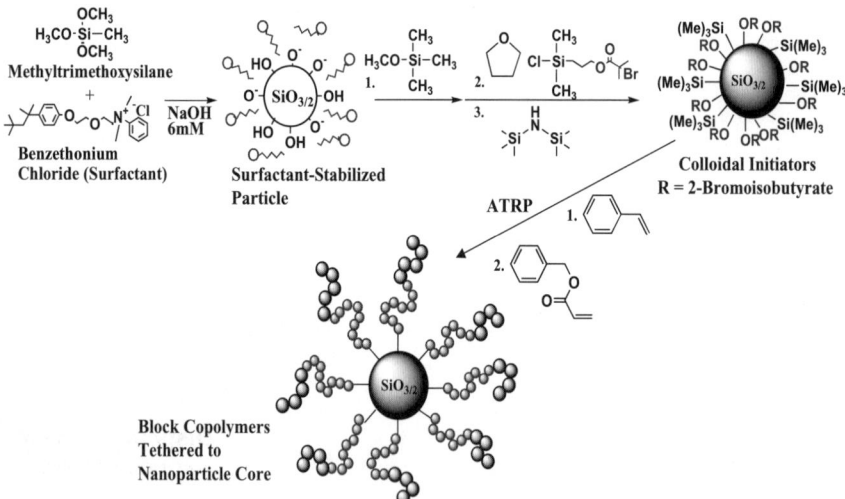

Scheme 54. Synthesis of hybrid nanoparticles using ATRP

polymerization of trialkoxysilanes [338]. Condensation reactions of the surface silanol groups with functional silanes yielded colloidal initiators bearing benzyl chloride, 2-bromopropionate, or 2-bromisobutyrate groups. These colloidal initiators were used in the ATRP of various vinyl monomers to produce the hybrid nanoparticles. Silica colloids with benzyl chloride groups on the surface were used for the ATRP of St. Dynamic light scattering (DLS) and transmission electron microscopy (TEM) confirmed that the diameter of the pSt hybrid nanoparticles increased with monomer conversion. GPC analysis of the pSt chains cleaved from the particles suggested polymer growth from the particle surface was controlled, as the polydispersities were low ($M_w/M_n<1.35$) and the experimental molecular weights agreed well with those predicted from the monomer conversion ($M_n=26{,}500$, $M_{n\,th}=30{,}600$). TEM images of pSt nanoparticle ultra-thin films revealed hexagonal packing of the colloids in a polymer matrix [336, 337, 339].

Farmer and Patten have modified the composition of the colloidal initiators to encapsulate cadmium sulfide particles in a shell of silica [340]. The ATRP initiating groups were introduced by condensing functional monoalkoxysilanes containing 2-bromopropionate groups to the silica surface. The ATRP of St from these core-shell colloidal initiators yielded an array of luminescent particles in a matrix of tethered pSt.

In a similar way, polysilsesquioxane colloids were functionalized with initiators for ATRP (cf. Scheme 54) [338]. Surface treatment of silanol groups with functional chlorosilanes possessing ~10^3 2-bromoisobutyrate groups, in addition to other silylating agents, yielded discrete colloids with ATRP initiating moieties. DLS and AFM both revealed that the particles were relatively uniform ($D_{eff\,DLS}=27$ nm, $D_{eff\,AFM}=19$ nm). The ATRP of St and benzyl acrylate (BzA) was then conducted to prepare hybrid nanoparticles with tethered block copolymers. GPC analysis of the cleaved pSt ($M_n=5250$; $M_w/M_n=1.22$) and pSt-b-pBzA ($M_n=27{,}280$; $M_w/M_n=1.48$) confirmed that the successive ATRPs of St and BzA from particles were successful. AFM tapping mode observations of the (sub)monolayers of the pSt-b-pBzA hybrid nanoparticles revealed the effect the tethered copolymer composition had on the morphology of the material on mica. In particular, AFM phase-contrast images showed that each component of the hybrid nanoparticle was discernible in ultrathin films cast onto mica (Fig. 48). These images implied that dark cores (colloidal initiators) were surrounded by a hard corona (tethered pSt segment) and dispersed within a light soft continuous matrix (tethered pBzA segment) [338]. Block copolymers were also grafted from functionalized silica nanoparticles [339].

Well-defined polymers have been attached to large particles ($D_{eff}>1$ μm) using CRP. Polymer coatings of controlled thickness and functionality were prepared using ATRP. In particular, hybrids from larger particles were synthesized as potential chromatographic stationary phases [324, 325] and templated supports [205, 341]. Huang and Wirth demonstrated the separation of various pro-

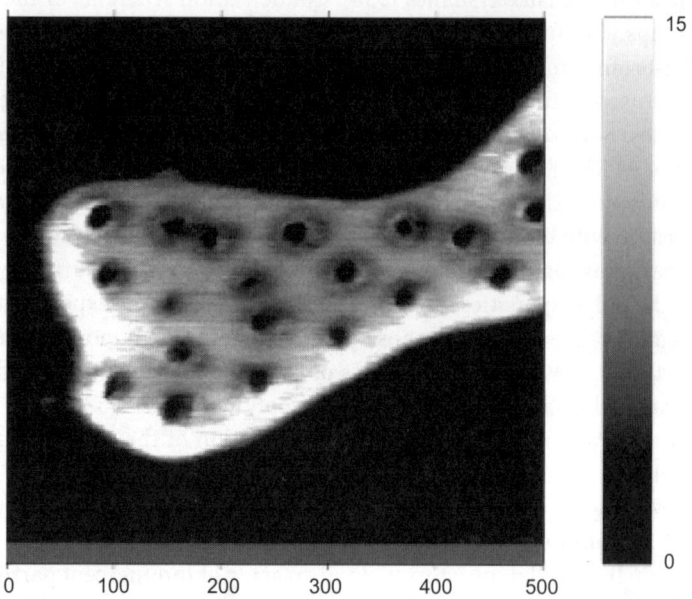

Fig. 48. AFM phase image of pSt-b-pBzA hybrid nanoparticle (sub)monolayer on mica. Domain assignments: dark spots (polysilsesquioxane particles), dark corona around particles (tethered pSt, M_n=5,250; M_w/M_n=1.22) light continuous matrix (tethered pSt-b-pBzA), M_n= 27,280; M_w/M_n=1.48). Reprinted with permission from [338]. Copyright (2001) American Chemical Society.

teins using poly (acrylamide) coated particles, which was evidence for the successful grafting of the polymers without significant clogging of the particle pores [342]. Similarly, Haddleton et al. used ATRP to grow polymethacrylates possessing nucleotide side chain groups from particles [341]. Oligonucleotide immobilization is an attractive approach for the templated synthesis of nucleic acids, with primary goals of controlling both the degree of polymerization and the sequence distribution in the final product. In a different templated system, Walt et al. grafted poly (benzyl acrylate) to a silica microsphere using ATRP, then treated the surface with hydrofluoric acid to prepare hollow polymeric colloids [343]. Hallensleban et al. also prepared hybrid particles with tethered pSt using ATRP from 2-chloro-2-phenylacetate functional silica particles [335].

Functional groups have also been placed on surfaces of organic colloidal particles prepared in an emulsion process. Vairon et al. applied ATRP to the homopolymerization of 2-hydroxyethyl acrylate (HEA) and 2-(methacryloyloxy)ethyl trimethylammonium chloride from the surface of a crosslinked polystyrene latex functionalized with alkyl bromide groups. ATRP was carried out using the surface groups of the dialyzed latex as the initiators. The resulting hydrophobic core/hydrophilic shell latexes were analyzed by FTIR, NMR, and DLS to confirm attachment of the hydrophilic polymer chains to the hydrophobic pSt particles [344].

4.1.5
Summary

Graft copolymers have been synthesized by CRP using three approaches: grafting from, grafting through, and grafting onto. Grafting from has perhaps been the most broadly utilized approach by both ATRP and nitroxide chemistries. Backbone macroinitiators have been prepared by modifying commercially available polymers, first prepared via free radical processes, metallocene chemistry, ionic mechanisms, and by CRP. An example of one extreme to which the grafting from technique can be taken is the synthesis of densely grafted copolymers, or polymer brushes, where there is a graft unit at every other carbon in the backbone, which was demonstrated using ATRP. The graft density can then be altered systematically by incorporating a comonomer that cannot be transformed into the ATRP initiating site, providing a route to tailor the properties of the brush to specific applications. A combination of ATRP with nitroxides can also be used for the grafting from approach because, although they are both CRP methods, the polymerizations occur under different conditions, allowing for an effective combination. The advantage to this approach is that both well-defined backbones and graft units can be prepared without any transformation chemistry.

Grafting through has been demonstrated using both nitroxide and ATRP approaches. One advantage to this approach is that well-defined graft units can be prepared, potentially through an entirely different polymerization mechanism, and then be incorporated into a backbone, again in a well-defined manner, using a controlled polymerization. A special feature of this approach is that, as the viscosity of the medium increases, the reactive chain end becomes more inaccessible due to diffusion effects. Incorporation also becomes more difficult as the molecular weight of the macromonomer increases, for similar reasons. This may result in a special gradient structure of the graft copolymers which have entirely new properties, different from regular grafts. While copolymerization of the macromonomers with small molecule monomers has been successful, only limited success with the homopolymerization of macromonomers to produce densely grafted structures has been realized. Subsequent development of dual living polymerization techniques has potentially made the grafting through methods obsolete (see below).

Surface modification is industrially an important application. Grafting from a substrate containing CRP initiating moieties can potentially result in the preparation of more uniform surfaces, as well as surfaces that can be modified to produce a desired effect, whether it be for microlithography applications or simply to impart a specific response to water. Although most of the surfaces studied thus far have been based on silica, growth from gold surfaces has recently been accomplished using ATRP. Due to the heat sensitivity of the alkyl thiols that link the polymers to the surface, nitroxide chemistry cannot currently be applied to gold surfaces because of the high temperatures required for their use. Further development of new nitroxides may eventually allow for their use in this application.

4.2
Star Polymers

Star-shaped polymers have been of interest since the development of CRP methods. Matyjaszewski et al. showed that ATRP can be used to prepare star polymers [114], when a hexafunctional initiator was used for the ATRP of St using the CuCl/bpy catalyst system, resulting in a 6-arm star with a narrow molecular weight distribution. Hawker et al. reported the synthesis of star polymers using TEMPO-mediated polymerizations. They demonstrated that star polymers of pSt could be prepared using a trifunctional unimolecular TEMPO initiator [294]. Since then, many reports on star polymers with various types of copolymer arms have been reported.

Matyjaszewski et al. used initiators based on functionalized inorganic compounds, whether derived from cyclotriphosphazene or cyclosiloxanes [232, 345]. The ATRP of St was carried out using 1,1, 3,3, 5,5-hexakis (4-bromomethylphenoxy) cyclotriphosphazene (Scheme 55) as the hexafunctional initiator and the CuBr/dNbpy catalyst at 105 °C in 50% diphenyl ether. The kinetic plot showed a constant concentration of active species throughout the polymerization, indicating termination was not significant and that the polymerization was well controlled. Deviation from linearity was observed in the plot of molecular weight vs conversion, but there was no evidence in the GPC traces to suggest this was due to loss of end-group functionality, so it was attributed to the hydrody-

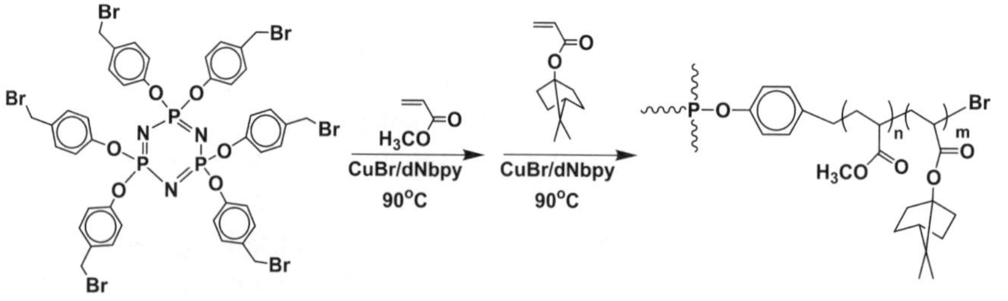

Scheme 55. The ATRP of MA and IBA initiated by 1,1,3,3,5,5-hexakis(4-bromomethylphenoxy) cyclotriphosphazene to produce an inorganic-organic star block copolymer [232, 345]

namic differences between the linear pSt standards and the star polymers [345]. This behavior was observed for the dendrimer graft copolymers as well [323]. One parameter that is vital to successful star formation is maintaining a low concentration of active species, thereby avoiding coupling reactions that may lead to gelation. Results from the ATRP of nBA with initiators containing 1, 2, 4, and 6 sites, but using the same concentrations of end-groups and catalyst (CuBr/dNbpy), showed that the rate of the polymerization was nearly identical in all the systems, as expected for systems with the same concentration of active species. The molecular weight evolution based on linear pSt standards, on the other hand, reflected the difference in the number of arms on the initiators, with deviation from the theoretical values becoming more pronounced as the number of arms increased. Light scattering analysis, however, produced good agreement between the experimental and theoretical molecular weight values. The molecular weight distributions were narrow ($M_w/M_n<1.2$), indicating good control over the polymerizations [345]. Furthermore, 6-arm star block copolymers with an initial pMA block were synthesized using 1,1,3,3,5,5-hexakis [4-(2-bromopropionyloxymethyl)phenoxy] cyclotriphosphazene and a CuBr/dNbpy catalyst at 90 °C ($M_n=20,000$, $M_w/M_n=1.14$, $M_{n, theo}=120,000$). This was chain extended with isobornyl acrylate (IBA) to produce a star block copolymer with $M_n=48,000$ and $M_w/M_n=1.37$. There was no evidence of coupling products nor unreacted macroinitiator in the GPC traces [232, 345].

Sawamoto et al. investigated the use of di- and trifunctional chloracetate initiators to carry out ATRP of MMA using the $RuCl_2(PPh_3)_3/Al(OiPr)_3$ catalyst system, then extended this to prepare calix [n] arene-based multifunctional cores [346, 347]. These cores were synthesized through the reaction of dichloroacetyl chloride and the corresponding calix [n] arene (Scheme 56) [346].

The initiators were characterized by ^1H NMR to confirm the structures were correct. No observable transesterification occurred with the multifunctional initiators and the Lewis acid. The experimental molecular weights for the polymers prepared from the tetrafunctional initiated reaction were lower than the

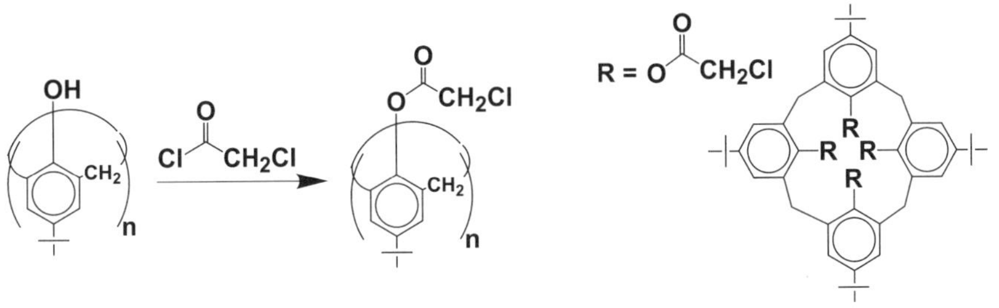

Scheme 56. Preparation of calixarene-core multifunctional initiators and the structure of calix [4] arene modified with acetyl chloride groups [346]

theoretical values but ^{1}H NMR analysis showed the true molecular weights were higher than those obtained by GPC [346]. The deviation stems from comparing star polymers to linear standards. To characterize the stars further, the arms were cleaved and analyzed by GPC. On average, two peaks were observed, one at the molecular weights corresponding to 1/4, 1/6, and 1/8 of that of the star polymer, and one corresponding to the original core initiator. The molecular weight distributions of the arms were significantly broader than those of the stars (M_w/M_n>1.5 vs <1.2), suggesting that the growth of the arms was not quite uniform and perhaps slow initiation occurred [346]. Nevertheless, an 8-arm pMMA star (M_n=27,100, M_w/M_n=1.14) was chain extended with nBMA under ATRP conditions to generate a star block copolymer with M_n=85,200 and M_w/M_n=1.10. There was little evidence of unreacted star macroinitiator, indicating that even if slow initiation occurred, the chain end functionality was maintained [346].

The ATRP of styrene from octafunctional 2-bromopropionate modified calixarenes was the focus of another study by Gnanou et al. [348]. Below 20% monomer conversion the polymerization was controlled based on the agreement between the measured and theoretical molecular weight values. Above that conversion, high molecular weight shoulders were observed by on-line light scattering measurements, which the authors attributed to star-star coupling. However, under the proper conditions of high dilution and cessation of the polymerization at low conversion, stars with molecular weights as high as M_n=340,000 were formed. There was agreement between the chain length of the isolated arm and the theoretically predicted values; however, no block copolymerizations were studied.

Hedrick et al. reported on the synthesis of 6- and 12-arm star-like block copolymers (Fig. 49) of tBA and MMA using sequential ATRP reactions with a $NiBr_2(PPh_3)_2$ catalyst [349]. The *tert*-butyl esters were subsequently deprotected

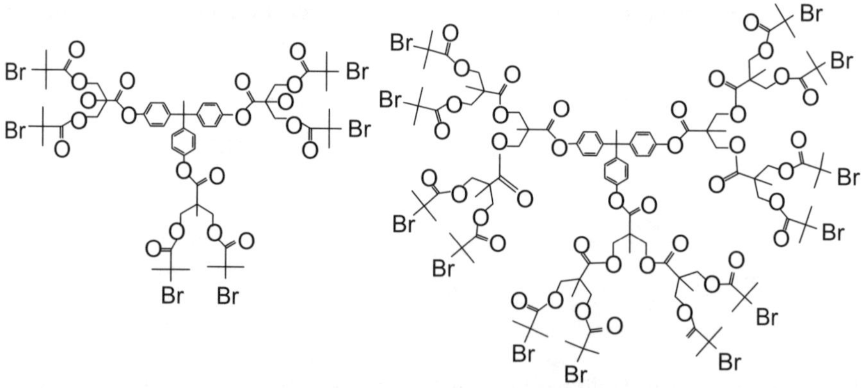

Fig. 49. Structures 6- and 12-arm ATRP initiators [349]

to yield amphiphilic stars. The ptBA stars were of molecular weight M_n=6800 (6 arms) and 16,500 (12 arms) with M_w/M_n=1.05 and 1.04, respectively. The clean chain extensions with MMA resulted in block copolymers of M_n=49,800 (M_w/M_n=1.29) and 74,500 (M_w/M_n=1.23) from both the short and long ptBA star initiators. The alternate order of block formation was also conducted; a 6-arm star of pMMA (M_n=15,000, M_w/M_n=1.13) was chain extended with tBA to yield a star block copolymer of M_n=22,000 with M_w/M_n=1.23 [349]. There was a high molecular weight shoulder in the pMMA star macroinitiator indicating some coupling had occurred during the polymerization; however, the GPC trace for the star block was symmetrical, indicating that it did not affect the chain extension reaction significantly. After deprotection, the ability of the stars to form micelles was investigated. In $CDCl_3$, the resonances from both blocks were visible; however, in going to a non-solvent for the pMMA, CD_3OD, the resonances corresponding to the pMMA segments were no longer visible. The opposite was true when acetone-d_6 was used; the poly(acrylic acid) resonances disappeared. This shows that micelles can indeed form and that the polarity of the solvent can affect their structure in solution [349].

Hedrick et al. also used dendritic 2-, 4-, 6-, and 12-arm multifunctional initiators for the ATRP of MMA [350] and for the copolymerization of MMA with HEMA [351]. The initiators were synthesized by the esterification of hydroxy-functional precursors with 2-bromo-2-methylpropionyl-functional units and were used for the ATRP of MMA using the $NiBr_2(PPh_3)_2$ catalyst system. Although the rates of the polymerizations were not discussed, the molecular weights of the star polymers containing various numbers of arms increased with increasing monomer: initiator ratios and showed little evidence of either unreacted initiator or products from coupling reactions [351]. ^1H NMR analysis appeared to confirm that all the initiating sites, regardless of the functionality, had participated in the polymerization; however, polymers derived from deuterated monomers indicated that for the 12-arm star, between 10 and 12 sites had actually initiated polymerization. The copolymerization of MMA and HEMA, carried out using 5–20% of HEMA relative to MMA, was successful using the 1-, 2-, 4-, and 6-arm stars. Although the molecular weight distributions were slightly broader than for the homopolymer MMA cases, they were M_w/M_n <1.4, indicating that the polymerization was still well controlled. These star copolymers were combined with methylsilsesquioxane (MSSQ) to investigate the compatibility of the two compounds as well as to determine whether the star copolymers could be used as pore generators for nanoporous thin films. The results indicated that the copolymer and MSSQ did not phase separate prior to curing, as was the case with the pMMA star homopolymers, and produced optically clear films. After thermal decomposition of the polymer, TEM micrographs showed voids in the MSSQ matrix of about 10 nm, thereby reducing the dielectric constant of the film and achieving the goal of the work [351].

Although no block copolymers have yet been reported, RAFT has also been used to prepare 4- and 6-arm pSt star polymers. Molecular weights ranged from M_n=26,000 to 80,000 with M_w/M_n=1.20–1.67. The rate of polymerization was slow however, reaching high conversion (>70%) only after 64 h [53]. The persistent radicals in RAFT may be trapped by growing chains resulting in the spontaneous formation of semi-stable three arm stars. This reaction in the presence of multifunctional initiators should lead to branching and crosslinking.

As illustrated in the block copolymer section, transformation techniques can be used to combine CRP methods with other living polymerization techniques. This technique has also been demonstrated for the synthesis of star copolymers. Hedrick et al. reported on the preparation of dendrimer-like star block copolymers using a transformation technique [352]. After using a hexahydroxy functional core as the initiator for ring-opening polymerization of CL, the hydroxy termini were esterified using various compounds to produce 6-, 12-, and 24-arm bromine based initiators. The transformation reaction was monitored by ^{1}H and ^{13}C NMR characterization. The initial 6-arm stars had M_n=14,300 and 56,000 with M_w/M_n=1.06 and 1.09, respectively [352]. The ATRP of MMA carried out using the $NiBr_2(PPh_3)_2$ catalyst system at 100 °C was successful, leading to the formation of block copolymers from all the initiators. The rate of polymerization was slow since the concentration of catalyst was deliberately kept low to prevent termination reactions; however, the GPC traces showed no evidence of unreacted macroinitiator in either chain extension reaction. The ATR copolymerization of MMA with HEMA or short pEO macromonomers was also successful, leading to polymers with various amphiphilic characteristics. Although the M_w/M_n<1.2, at higher catalyst ratios there was some evidence of coupling in the products [352]. The thermal characteristics of the star, and dendrimer-like stars, were dependent on the composition and the number of arms. Only single phase transitions were evident as the number of arms increased in the pCL-*b*-pMMA copolymers; however, when the pEO was randomly incorporated into the pMMA block, two transitions were observed in the polymers with more arms [352].

Both Xu and Pan [353] and Kennedy et al. [354] prepared star cores via living cationic polymerization, then transformed the end groups to ATRP initiators to make star block copolymers. Xu and Pan used a tetrafunctional initiator for the cationic ring-opening polymerization of THF, then esterified it with bromoacetyl or bromoisobutyryl chlorides (Scheme 57).

Using the primary bromoacetyl-based pTHF macroinitiator (M_n=2100, M_w/M_n=1.20) with a CuBr/bpy catalyst for the ATRP of St resulted in poor initiation efficiency (~50%) and broad molecular weight distributions [353]. However, when the tertiary bromoisobutyryl-based pTHF macroinitiator was employed (M_n=1700, M_w/M_n=1.25), the kinetic plot had little deviation from linearity to suggest either slow initiation or termination and the molecular weights increased

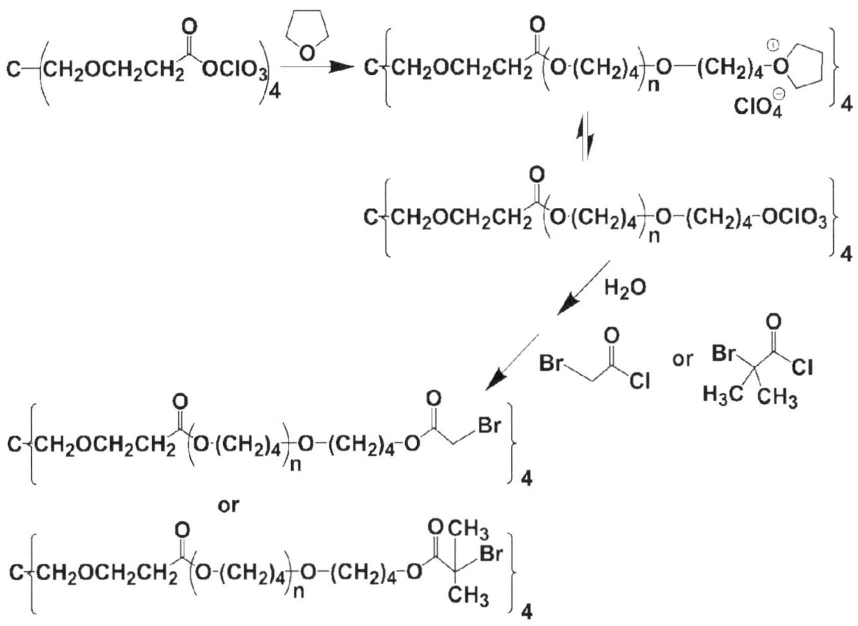

4 Other Chain Architectures 143

Scheme 57. Synthesis of pTHF via CROP followed by transformation to an ATRP macroinitiator [353]

linearly with conversion and had narrow molecular weight distributions ($M_w/M_n<1.2$). Cleaving the arms resulted in polymers with approximately 1/4 the molecular weight of the star polymer, suggesting that the polymerization was well-controlled ($M_w/M_n<1.25$). The (pTHF-b-pSt-Br)$_4$ macroinitiator was used to initiate the bulk ATRP of MMA, again using the CuBr/bpy catalyst, at a monomer:initiator ratio of 3200:1 with a fourfold excess of catalyst relative to the concentration of initiator. The molecular weight of the macroinitiator increased from $M_n=28,200$ to $M_n=183,000$ and the molecular weight distribution decreased from $M_w/M_n=1.30$ to $M_w/M_n=1.18$. There was a clean shift of the GPC trace to higher molecular weights, indicating that initiation was efficient and termination was negligible [353].

Kennedy et al. used living cationic polymerization from a tricumyl initiator to prepare an allyl-terminated 3-arm star of pIB, followed by hydroboration/oxidation to generate hydroxy chain ends which were esterified with 2-bromoisobutyryl bromide to generate the ATRP trifunctional macroinitiator (Scheme 58) [354]. They subsequently carried out ATRP of MMA in toluene using the Cu(I)/N-(n-pentyl)2-pyridylmethanimine catalyst system with the addition of Cu^0 powder [242] to maintain a sufficient concentration of active Cu(I) [354]. Macroinitiators of $M_n=9200$ and 15,000 were prepared and both had narrow molecular weight distributions ($M_w/M_n=1.15$ and 1.09, respectively). The formation of block copoly-

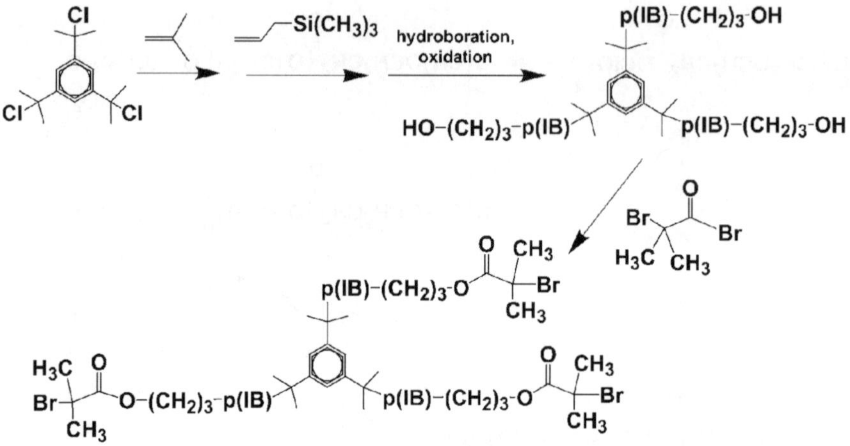

Scheme 58. Living cationic polymerization of isobutylane followed by allylation, hydroboration/oxidation, and esterification with 2-bromoisobutyryl bromide to generate the three-armed ATRP macroinitiator [354]

mers with MMA, using either macroinitiator, were well-controlled, resulting in the preparation of star-block copolymers with narrow molecular weight distributions and molecular weights that surprisingly agreed with the theoretical values [354].

Using CuCl instead of CuBr to form the catalyst for the chain extension reactions produced polymers with lower polydispersities, as expected when halogen exchange is employed [175]. Core destruction provided information regarding the block copolymer arms. After oxidative degradation, the diblock arms of a star block with M_n=30,900 and M_w/M_n=1.50 had M_n=10,600 with M_w/M_n=1.89, suggesting efficient initiation from all the sites. Treating the star-block with a strong base led to dehalogenation of the end groups, producing olefinic end groups as well as to the hydrolysis of the methyl esters to methacrylic acid groups, making an amphiphilic star-block copolymer. Thermal and mechanical analysis supported the formation of block copolymers, as did observation of two glass transition temperatures for all stars and the fact that the tensile strength of the stars increased as the length of the pMMA segment increased [354].

Gnanou et al. used the living anionic ROP of EO to generate hydroxy-terminated pEO cores of various functionalities. These were subsequently transformed into ATRP initiators via an esterification reaction with 2-bromopropionyl bromide (M_n=2200 to 20,000, M_w/M_n=1.07 to 1.10) [355]. The complete transformation was confirmed by ^{1}H NMR analysis. Bulk ATRP of St using 3- and 4-arm stars was carried out using the CuBr/bpy catalyst system at 100 °C, but was kept to low monomer conversions to avoid the coupling reactions observed previously [348]. Clean chain extension was achieved with significant increases in the molecular weights of the pEO cores and polymers exhibited molecular weight distributions $M_w/M_n<1.3$. After destruction of a 3-arm star block using KOH, the molecular weights of the pSt segments determined by ^{1}H NMR (M_n=8000) agreed well with those found from GPC analysis (M_n=10,000), indicating that the ATRP chain extension reaction was well-controlled. Initiators containing higher functionalities were obtained by using acid chlorides that had multiple bromoesters in the esterification step. This led to $(pEO)_n$-b-$(pSt)_{2n}$ star blocks where n=1, 2, 3, or 4, producing a more dendritic-type structure. After carrying out the ATRP of St using conditions similar to the 3- and 4-arm star reactions, the resulting polymers ranged in molecular weight from M_n=7600 to 123,000 (as determined by ^{1}H NMR) with narrow molecular weight distributions ($M_w/M_n<1.40$) [355]. GPC analysis was not useful for molecular weight determination since the results varied as a function of concentration. Aggregation of the polymers into micelles was observed, particularly with those composed of a larger number of arms. The ability of the polymers to form micelles was confirmed by ^{1}H NMR characterization performed in selective solvents. In $CDCl_3$, the resonances corresponding to both polymers were visible; however, when CD_3OD was added, the pSt resonances disappeared, suggesting that the hydrophilic pEO surrounded a hydrophobic pSt core [355]. This approach to preparing multi-arm polymers may provide yet another route for tailoring the properties of amphiphilic copolymers to particular applications.

All of the aforementioned reports showed star polymer formation originating from a core. The so-called "arm-first" approach has also been demonstrated.

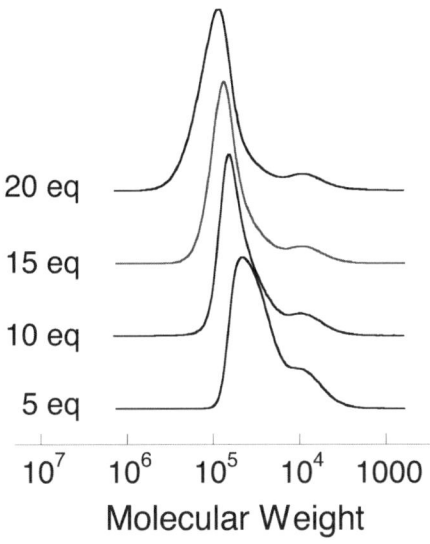

Fig. 50. GPC traces of a ptBA arm and the corresponding star polymer after coupling with DVB. Reprinted with permission from [357]. Copyright (2000) American Chemical Society.

Linear polymers of polystyrene [356] or poly(*tert*-butyl acrylate) [357] were first prepared by ATRP. The resulting polymers were subsequently allowed to react with a crosslinking reagents such as divinyl benzene, 1,4-butanediol diacrylate, or ethylene glycol dimethacrylate to form crosslinked cores. Several factors pertinent to star polymer formation, including the choice of the exchanging halogen and solvent, the addition of a copper(II) species, the ratio of the coupling reagent to the macroinitiator, and the reaction time for the star formation are crucial for efficient star formation. The highest efficiency (~95%) was observed with 10- to 15-fold excess of the difunctional monomer over chain ends (Fig. 50). Functional initiators were used to directly prepare arms with α-functionalities since ATRP is highly tolerant to functional groups. End-functional star polymers with hydroxy, epoxy, amino, cyano, and bromine groups on the outer layers were successfully synthesized [357]. An alternative approach to end-functional stars can employ a chain end transformation process, such as a radical addition

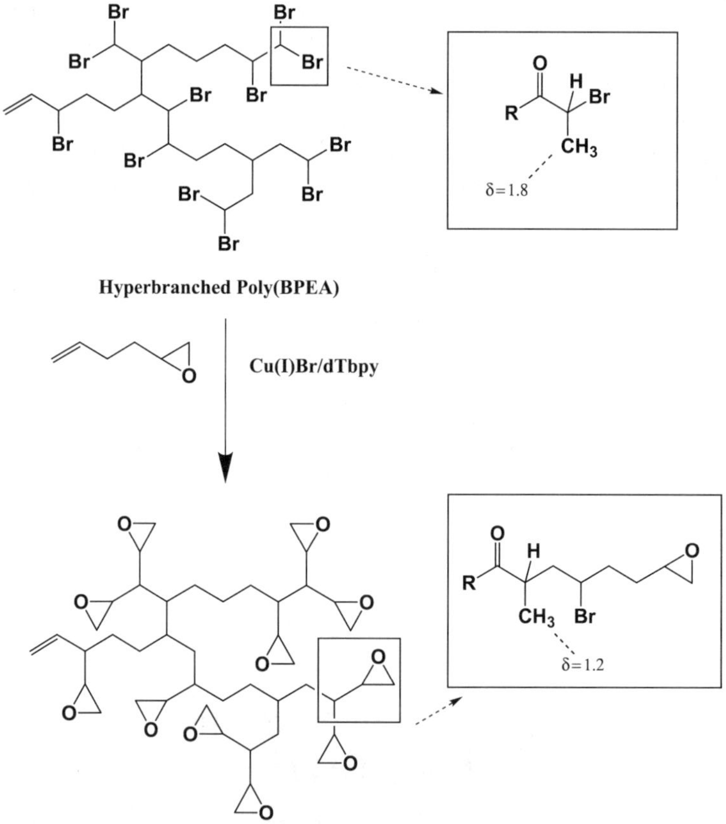

Scheme 59. Addition of 1,2-epoxy-5-hexene to the hyperbranched pBPEA [89]

reaction to incorporate epoxy or hydroxy groups [89]. Since the core of these stars contains a segment other than the shall, they can be considered block copolymers.

Hyperbranched polymers have also been used as cores for star polymers. Up to 80 arms of pBA were grown from the core of pBPEA with DP~80. Since number of functional groups is equal to the degree of polymerization, incorporation of just one unit per chain end brings the content of another comonomer to ~50% [89]. Hyperbranched polymers prepared by ATRP were modified in just such a way using monomers that were non-polymerizable by ATRP but carried useful functionalities such as epoxies or alkenes, as shown in Scheme 59.

4.3
Simultaneous/Dual Living Polymerizations

Perhaps one of the most exciting developments in the field of CRP has been the synthesis of polymers from initiators that contain sites from which two (or more) different polymerization methods can be carried out. The tolerance of radicals to various functionalities and a broad range of polymerization conditions have allowed use of these multifunctional initiators for controlled polymerization reactions to be successful. In 1997, Puts and Sogah reported on the synthesis of an orthogonal initiator that contained three sites: a TEMPO moiety for CRP, a protected hydroxyl group for anionic ring-opening polymerization, and an oxazoline unit that could either be used for the cationic ring opening polymerization of oxazoline or, since the position α to the oxazoline is acidic, for an anionic vinyl polymerization [358]. The structure of the initiator is shown in Fig. 51. Initially, the CRP of St was carried out by heating the initiator with St at

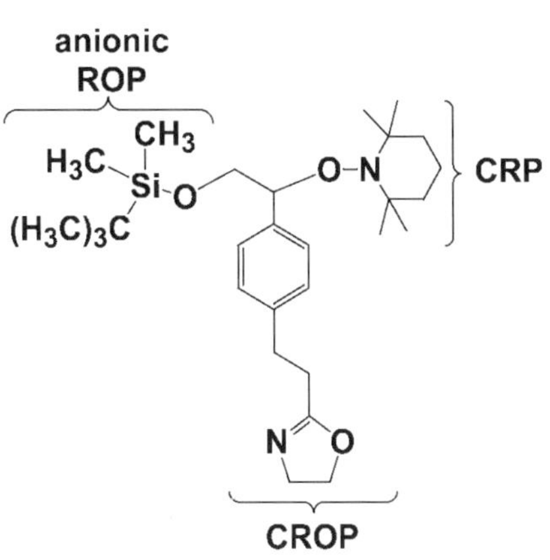

Fig. 51. Structure of an orthogonal initiator containing sites for CRP, anionic ROP, and CROP [358]

135 °C, forming a macroinitiator (M_n=5000, M_w/M_n=1.16). The integrity of the remaining sites was confirmed by ^{1}H and ^{13}C NMR characterizations. The polymer was then treated as a macromonomer and the cationic ring opening polymerization of oxazoline initiated by methyl triflate was carried out at 140 °C to generate a graft copolymer. Although the rate of the homopolymerization was slow, GPC analysis of the products showed that the trace corresponding to the macromonomer decreased significantly, and a peak corresponding to the graft copolymer appeared. ^{1}H and ^{13}C NMR indicated that the protecting group on the hydroxyl moiety remained after the cationic polymerization step [358].

Later work with this same initiator by Sogah et al. showed that the CRP of St and the cationic ring opening polymerization of phenyl oxazoline (PO) could be accomplished simultaneously, producing block copolymers instead of graft copolymers [359]. The polymerizations were successful at various feed ratios and monomer:initiator ratios, producing block copolymers with M_n=24,300–33,500 and M_w/M_n=1.27–1.40. Homopolymer pSt contaminants were not present. Hydrolysis of the oxazoline residue resulted in formation of a water-soluble poly(ethylene imine) block. After dialysis, GPC analysis showed no homopolymer contaminants, again supporting the conclusion that the pSt segments were connected to the pPO segments [359]. The amphiphilic block copolymer showed evidence of micelle formation when selective solvents were used for ^{1}H NMR analysis, as the pSt resonances disappeared when CD_3OD was used as the solvent. Thermal analysis indicated the presence of a glass transition temperature for the pSt segment and a melt transition for the pPO. The simultaneous anionic ring-opening polymerization of CL and the CRP of St was also successfully carried out [359].

Another series of papers has focused on combining both ATRP and nitroxide-mediated polymerizations with condensation and ring-opening polymerization reactions [360–364]. Initial reports by Hawker et al. [360, 365] and Jerome et al. [361] used concepts similar to those first put forth by Puts and Sogah and prepared initiators that were dual-headed and could be used for two different polymerization techniques without a transformation step. They found that identification and use of the proper conditions could allow for the simultaneous polymerizations of two different monomers by two different routes.

Using a unimolecular TEMPO initiator containing a hydroxy functionality, the simultaneous ring-opening polymerization and controlled radical polymerizations of CL (using $Sn(Oct)_2$ as the catalyst) and St were carried out (Scheme 60) [362]. The molecular weights increased linearly with conversion, indicating the living nature of both processes.

A range of monomer feed compositions were used, resulting in polymers with molecular weights ranging from M_n=6300 to 21,000 with M_w/M_n<1.8 and pCL contents ranging from 0.26 to 0.67 [362]. The simultaneous ATRP of MMA using the $NiBr_2(PPh_3)_3$ catalyst system and ring-opening polymerization of CL using

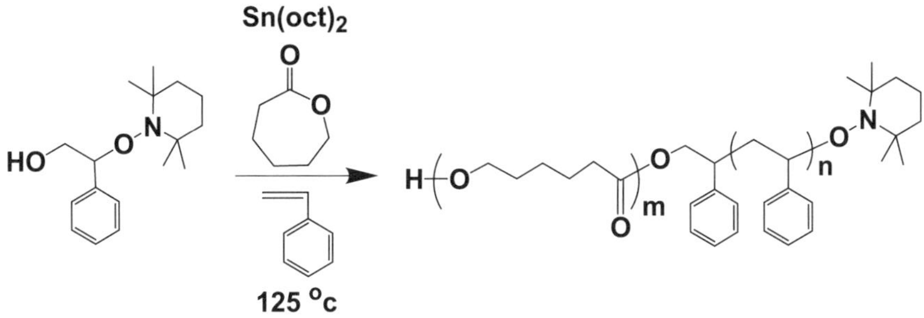

Scheme 60. Simultaneous ROP of caprolactone and TEMPO-mediated CRP of St [362]

Al(OiPr)$_3$ was also achieved using 2,2,2-tribromoethanol as the initiator for both processes. Again, the molecular weight control was good; however, the molecular weight distributions were broader than in the prior polymerizations. Molecular weight distributions were narrower in the presence of pyridine, which limited any transesterification that might occur with the MMA and Al(OiPr)$_3$ [362]. Thermal analysis showed the presence of two glass transitions in the pCL-b-pSt copolymer. Hydrolysis of the pCL-b-pMMA block copolymer resulted in a significant decrease in the molecular weight (M_n=70,000 to 26,000) and the molecular weight distribution (M_w/M_n=1.4 to 1.2). Both served to confirm that block copolymers were indeed prepared. To illustrate that more complex architectures could be synthesized, MMA was copolymerized with HEMA using ATRP with the RhCl(PPh$_3$)$_3$ catalyst while the hydroxy functionality on the HEMA was used to initiate the ring-opening polymerization of CL in the presence of Al(OiPr). When a ratio of 1:0.1:1 of MMA, HEMA, and CL was used the polymerization reached 73% total monomer conversion after 18 h at 50 °C. ^{1}H NMR confirmed the presence of both pMMA and pCL; after hydrolysis of the pCL graft units, the disappearance of which was confirmed using ^{1}H NMR, the backbone had M_n= 30,000 with M_w/M_n=1.25 [362].

Miller et al. combined condensation polymerizations with both nitroxide-mediated and ROP (Scheme 61) [363]. The condensation polymerization of 2,7-dibromo-9,9-dihexylfluorenes (DHF) was carried out in the presence of a Ni(COD)$_2$ catalyst using (1-(4'-bromophenyl)1-(2'', 2'', 6'', 6''-tetramethyl-1-piperidinyl-oxy)ethyl) as a capping agent/initiator for the CRP of St. This led to a central block of pDHF and outer blocks of pSt (M_n=36,000, M_w/M_n=1.52). Similarly, a diol functionalized with a bromine moiety was used as the end capper/initiator for the simultaneous polymerizations of CL and DHF in the presence of both Ni(COD)$_2$ and Sn(Oct)$_2$ (M_n=17,000–31,600, M_w/M_n=1.65–2.39) [363]. The molecular weight distribution narrowed as the content of the DHF in the feed decreased for the condensation/ROP combination. Combining ATRP

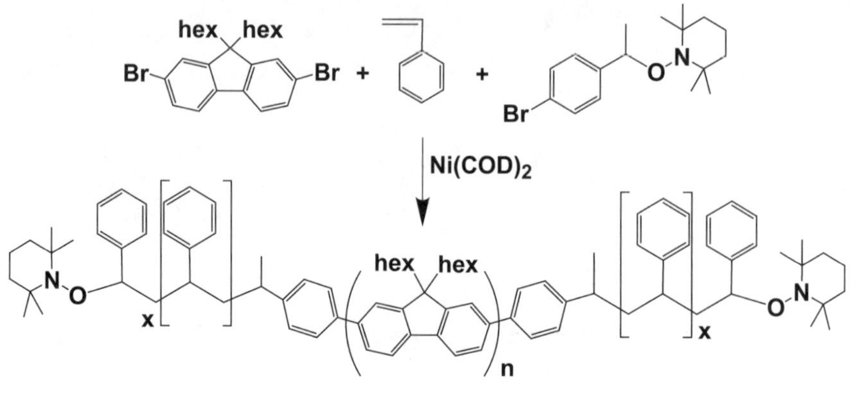

Scheme 61. Simultaneous condensation and TEMPO-mediated polymerizations [363]

with the condensation polymerization was only successful in a stepwise manner, not a simultaneous one, which was attributed to the instability of the α-haloesters under the conditions needed for the condensation polymerization (M_n= 83,000, M_w/M_n=1.64). Formation of the block copolymers was confirmed by ^{1}H NMR analysis both before and after hydrolysis of the pCL from the pDHF [363].

As detailed previously, Grubbs et al. have combined ROMP with ATRP to produce block copolymers [293]. In a one-pot reaction, ROMP, ATRP and hydrogenation was carried out successfully.

Hedrick et al. have introduced the concept of a monomer that can function as an ATRP initiator, as well as participate in a polymerization proceeding by an alternate mechanism [364]. The monomer/initiator was used to prepare graft copolymers via the grafting from, grafting through, and simultaneous living polymerization routes (Fig. 52). γ-(2-Bromo-2-methylpropionyl)-ε-caprolactone was homo- and co-polymerized with CL to yield a macroinitiator with a variable number of grafting sites for the ATRP of MMA. The composition of the backbone was predictable based on the monomer feed ratio, as were the molecular weights, with molecular weight distributions M_w/M_n<1.6. The ATRP of MMA was carried out using the $NiBr_2(PPh_3)_2$ catalyst system. The graft copolymers had compositions that were predictable based on the molar feed ratios, predictable molecular weights (M_n=35,000–47,000) and narrow molecular weight distributions (M_w/M_n=1.28–1.41) [364]. Cleaving the arms with HCl resulted in polymers with M_n=13,600 and M_w/M_n=1.06, confirming that the grafting procedure had been controlled. Conversely, the ATRP of MMA can be carried out first to yield a CL-capped pMMA macromonomer that can be polymerized via ROP. The molecular weights of the macromonomers ranged from M_n=2200 to 3800 with M_w/M_n=1.15–1.22. Using $Sn(oct)_2$ as the catalyst, these macromonomers

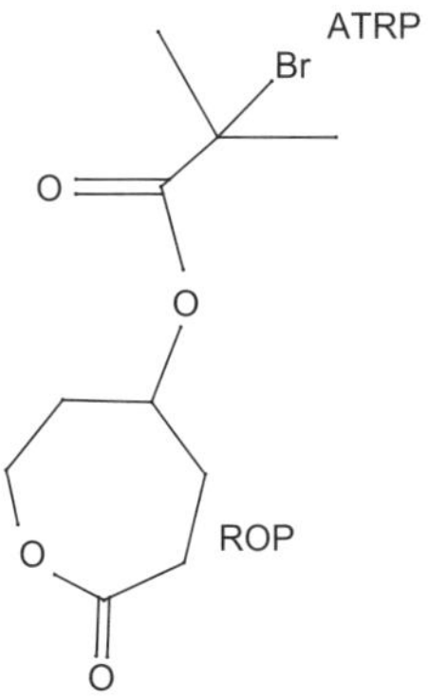

4 Other Chain Architectures 151

Fig. 52. Structure of γ-(2- bromo-2-methylpropionyl)-ε-caprolactone [36]

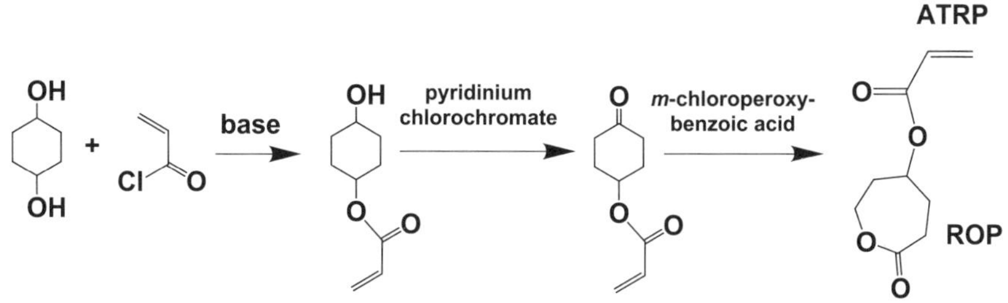

Scheme 62. Synthesis of 4-(acryloyloxy)-e-caprolactone [366]

were copolymerized with CL to prepare the graft copolymers in high yield (>80%). The mole fractions of macromonomer in the copolymers matched well with the values predicted by the comonomer feed ratios. The GPC traces moved cleanly to higher molecular weights. The final approach to such graft copolymers was simultaneous ROP and ATRP reactions. The final polymer had $M_n=$ 12,000 with $M_w/M_n=2.30$, suggesting that the polymerization was not as well controlled as for the other approaches. No other characterization information was provided [364].

Subsequently, Hedrick et al. also prepared a monomer that contains two different reactive sites, allowing for the polymerization of a single monomer by two different mechanisms [366]. The synthesis of the monomer is shown in Scheme 62. The 4-(acryloyloxy)-ε-caprolactone was then polymerized via an ATRP mechanism with $NiBr_2(PPh_3)_2$ as the catalyst at 90 °C to yield polymers with $M_n=3500-24,000$ with $M_w/M_n=1.13-1.30$.

The glass transition temperature of the homopolymers was 95 °C. Alternatively, the ROP of the CL portion using Al(OiPr)$_3$ as the catalyst at 25 °C resulted in homopolymers with M_n=1800–14,000 and M_w/M_n=1.15–1.22 [366]. 6-Arm star polymers composed of random copolymers of 4-(acryloyloxy)-ε-caprolactone with CL or L,L-lactide were also prepared, although the control over the polymerization was lost in the presence of L,L-lactide. These two approaches to polymerizing the same monomer led to polymers with novel pendant functionalities that may be useful for a wide variety of applications [366].

5
Overall Summary

5.1
General Overview

This review has summarized the information available in the literature regarding various copolymers that involved some aspect of controlled/living radical polymerization (CRP) for their synthesis. The topics addressed included the formation of statistical, periodic and gradient copolymers, block and graft copolymers, including tethered copolymers, and segmented polymers with star-type chain architectures. There are numerous examples of the synthesis of all of these types of copolymers using only CRP techniques, whether it is the use of nitroxide-mediated systems, ATRP, or the degenerative transfer approach. In addition, other polymerization methods have been combined with CRP to afford segmented copolymers whose compositional combinations were never before possible. Living carbocationic, carbanionic, various types of ring-opening polymerization, coordination polymerization, and even step-growth processes have all been used in conjunction with CRP methods to produce polymers with novel structures. Finally, one of the most unique approaches to the preparation of segmented copolymers has been the development of initiators and monomers that can be used for two types of polymerization processes, whether it be in a stepwise or concurrent manner. These types of advances lend themselves to the development of routes for the preparation of materials with properties that can be specifically tuned for particular applications.

5.2
Critical Evaluation of CRP Methods for Materials Preparation

All the CRP methods have strengths that can be exploited in particular systems. TEMPO is essentially useful only for the polymerization of styrene-based monomers, whether for the preparation of statistical or block copolymers [38]. The radicals generated through the self-initiation of St help to moderate the rate of polymerization by consuming any excess TEMPO generated by termination reactions, which will not occur with other monomers. Acrylate monomers, for example, are very sensitive to the concentration of free TEMPO and therefore its build-up causes the polymerization to stop. The use of different nitroxides and alkoxyamines like DEPN [73] and TMPAH [71], which provide higher equilibrium constants and allow for faster polymerization rates, has also enabled the homo- and copolymerizations of acrylate monomers, as well as for St at lower temperatures. Block order is important, however, and chain end functionality is reduced when TMPAH functional polymers are chain extended with BA. This may place limits on its applicability for materials synthesis. Various copolymers have been prepared using TMPAH, including those of St with MMA, acrylic acid, 2-hydroxyethyl acrylate, glycidyl acrylate, and acrylamides. It has also been used successfully for polymerizing isoprene, which can additionally be incorporated into statistical and block copolymers. However, methacrylate monomers still cannot be homopolymerized in a controlled fashion [165]. This is a serious drawback for many industrial applications because it eliminates a whole class of monomers with desirable properties.

The major strength of ATRP is that not only can the concentration of catalyst be adjusted to increase or decrease the activity of particular system, but the nature of the catalyst can be altered as well [49, 367]. Polymerization temperatures can range anywhere from below room temperature (acrylates with the Cu-Br/Me$_6$TREN system) to above 130 °C (styrene with a CuCl/bpy catalyst) [42, 172]. The dormant chain end groups on the polymers are halogens and simple substitution or addition chemistry can be used either to remove them or to transform them to other useful functionalities, like azides, amines, hydroxides, or double bonds [87]. Acrylate-, methacrylate-, acrylamide-, and styrene-based monomers have all been polymerized using ATRP, and the molecular weights for linear polymers range from several hundred to several hundred thousand. Varying the block order is simple, since the halogen exchange technique allows for very good control over the polymerization of methacrylates when less reactive macroinitiators are utilized. ATRP may be particularly useful for the synthesis of such "difficult" block copolymers, those with relatively low molar mass, i.e., high content of end groups which are least expensive from all CRP methods, and for various hybrid materials which require efficient incorporation of many initiating sites [181]. ATRP methods have as a limitation the fact that a catalyst is needed and therefore has to be recycled or removed [368]. Neither α-olefins nor vinyl acetate have yet been polymerized using ATRP in a well controlled manner [221], but with further catalyst development this is expected to be achieved.

Degenerative transfer, including RAFT, is useful for a variety of monomers, from methacrylates to vinyl acetate [54, 55]. However, the concept of degenerative chain transfer requires a supply of additional radicals (from initiator, light, or a thermal process) which limits blocking efficiency and end-functionality. In addition, some dithioesters may act as retarders and, as the concentration of chain transfer agent increases, the rate of polymerization decreases, leading to longer reaction times for lower molecular polymers. Trapping the persistent radicals in RAFT with growing chains may lead to branching and crosslinking when multifunctional initiators are used. Generally, polymers with chain ends that have lower transfer rates (i.e., methacrylates) must be polymerized prior to monomers with higher transfer rates (i.e., styrenes and acrylates) in order to have high blocking efficiency. Block order is therefore critical to material synthesis. While this is also a problem in ATRP, halogen exchange can be used to overcome it. As demonstrated with ATRP and the nitroxides, however, new developments are constantly improving the systems and allowing limitations to be overcome. This may likely occur in the RAFT systems as well.

All CRP chemistries should be useful in dual living polymerization techniques directed at the preparation of graft and block copolymers. However, interference may sometimes occur between, e.g., ROP catalyst and either ATRP catalyst, or end functionalities in NMP or RAFT systems.

5.3
Potential Applications for Copolymers Made by CRP Methods

Copolymers prepared using CRP methods can be utilized for multiple applications. To optimize the targeted use, it is possible to vary not only the copolymer composition and properties of individual comonomers, but also the chain topology and end functionalities. The range of monomers polymerizable by CRP is similar to conventional systems, though NMP and ATRP have some limitations. Polymers from very non-polar monomers such as dienes and styrenes or very polar ones like acrylates or even acrylic acids can be prepared. (Meth)acrylate monomers with lipophobic fluoroalkyl groups, hydrophilic oligo(EO), hydroxyethyl, dimethylaminoethyl, or acidic moieties, and even hydrophobic lauryl and stearyl groups represent the different properties individual monomers within the same family can have. Additionally, crystallization (for long alkyl groups), high and low T_g (isobornyl and ethylhexyl) help to manipulate thermal and mechanical properties. Statistical copolymerization of two or more comonomers with different structures was previously used in conventional radical polymerizations to fine tune many properties. However, as explained before, this often resulted in differences between the polymer chains produced early and late in the process, which could produce an inhomogeneity in the system. Because in CRP all the polymer chains start growing at the same time, the differences in the monomer feed are reflected in each chain, resulting in gradient copolymers. These copolymers have properties that are different from conventional random copolymers and segmented copolymers such as blocks and grafts. They may phase separate, but the shape of gradient allows for control and tuning of the order-disorder temperature and other properties.

CRP, as with any living system, can be used to prepare block copolymers. One application may be their use as thermoplastic elastomers. This is the case for ABA triblock copolymers composed of soft pIB inner blocks and hard pSt outer blocks [269], as well as soft pnBA middle segments and hard pMMA outer segments, whether containing only "clean" blocks or containing blocks with a gradient at the block interface [91, 94, 177, 179]. The latter, due to a broad range of transitions, may also be excellent adhesives. Perhaps even more interesting is to combine incompatible segments that lead to amphiphilic systems which can be used as surfactants. This was shown for a variety of pSt-poly (acrylic acid) copolymers with several different chain topologies and architectures that were prepared by ATRP and utilized as surfactants for the emulsion polymerization of St [194, 195]. Another potential application is to use double hydrophilic block copolymers for crystal engineering.

The segments may be connected not only in linear fashion but also in a comblike structure to form graft copolymers. Graft copolymers are generally easier to make than block copolymers and tolerate more errors. Grafts can be made by grafting from, through, and onto. In addition, it is possible to change the grafting density along the chain to form gradient-like grafts with properties that differ dramatically from regular graft copolymers. Grafting can also be used to modify the properties of some commodity polymers such as polyolefins or pVC. Commercial pVC contains additives that act as plasticizers and impart elasticity to the material. Unfortunately, these additives can leach out of the material. A homogeneous graft copolymer of pVC with nBA acts as a self-plasticizing material, eliminating the need for additives [297]. This may be desirable for applications such as children's teething rings and feeding bottles as well as for biomedical applications.

Grafting from surfaces using CRP techniques has allowed for precise patterning of the surfaces as well as tuning of the surface properties for specific microlithographic applications [332, 369]. An additional possibility may be control over the materials properties and the ability to tune them for particular application by incorporating functionalities into the polymer chain. They can be used for specific interactions with surfaces, to enhance compatibility, for the reactive processing and blending, or be crosslinked, as for coatings.

Polymers prepared via CRP show promise for applications like photoresists [112], liquid-crystalline displays [147–149, 154], and photocatalysts [151]. Incorporating blocks prepared using CRP techniques into copolymers with conductive or luminescent blocks [240, 241, 243, 251] may impart better processability and make them useful for a broader range of applications. Block or gradient copolymers with highly controlled compositions may also be industrially useful as blend compatibilizers or as surfactants [194], perhaps improving upon already existing materials. Well-defined or functional compatibilizers and stabilizers could potentially result in lower production costs if less material is needed to impart the desired properties.

The ability to control polymer structure so completely using CRP will allow one to develop a more comprehensive structure/property correlation [370]. This has been done for several systems but requires much more systematic investigation. The ease of manipulating the fundamental characteristics of polymers (molecular weight, chain topology, chain architecture, and compositional) prepared using CRP methods may make development of new, performance materials targeting specific applications much more economical as the advantages of all the possibilities offered by CRP for polymerization of available monomers is exploited.

Acknowledgement. ATRP research carried out at Carnegie Mellon University was supported by the National Science Foundation, Environmental Protection Agency as well as ATRP and CRP Consortia at CMU (Akzo, Asahi, Atofina, Bayer, BFGoodrich, BYK, Cabot, Ciba, DSM, Elf, Geon, GIRSA, JSR, Kaneka, Mitsubishi, Mitsui, Motorola, 3M, Nalco, Nippon Goshei, Nitto Denko, PPG, Rohm & Haas, Rohmax, Sasol, Solvay, Teijin and Zeon). The success of this work has largely depended on creativity and devotion of many researchers whose names are listed in references.

References

1. Bamford CH, Barb WG, Jenkins AD, Onyon PF (1958) The kinetics of vinyl polymerization by radical mechanisms. Academic Press, New York
2. Bagdasarian HS (1959) Theory of radical polymerization. Izd Akademii Nauk, Moscow
3. Moad G, Solomon DH (1995) The chemistry of free radical polymerization. Elsevier Science, Bath
4. a) Matyjaszewski K, Gaynor SG (2000) In: Craver CD, Carraher CE Jr (eds) Applied polymer science. Pergamon Press, Oxford, UK, p 929, b) Matyjaszewski K, Davis TP (2002) Handbook of Radical Polymerization, Wiley-Interscience
5. Greenley RZ (1999) In: Brandrup J, Immergut EH, Grulke EA (eds) Polymer handbook. Wiley, New York, p II
6. Greszta D, Matyjaszewski K (1996) Polym Prepr (Am Chem Soc, Div Polym Chem) 37:569
7. Pakula T, Matyjaszewski K (1996) Macromol Theory Simulat 5:987
8. Matyjaszewski K, Ziegler MJ, Arehart SV, Greszta D, Pakula T (2000) J Phys Org Chem 13:775
9. Szwarc M (1956) Nature 178:1168
10. Webster OW (1991) Science 251:887
11. Penczek S, Kubisa P, Matyjaszewski K (1985) Adv Polym Sci 68/69:1
12. Penczek S, Matyjaszewski K (1976) J Polym Sci, Polym Symp 56:255
13. Kennedy JP, Ivan B (1992) Designed polymers by carbocationic macromolecular engineering. Theory and practice. Hanser, Munich
14. Sawamoto M (1991) Progr Polym Sci 16:111
15. Matyjaszewski K (1996) (ed) Cationic polymerizations: mechanisms, synthesis, and applications. Marcel Dekker, New York
16. Matyjaszewski K, Sigwalt P (1994) Polymer Int 35:1
17. Greszta D, Mardare D, Matyjaszewski K (1994) Macromolecules 27:638
18. Matyjaszewski K (1996) Curr Opin Solid State Mater Sci 1:769
19. Otsu T, Matsumoto A (1998) Adv Polym Sci 136:75
20. Matyjaszewski K, Müller AHE (1997) Polym Prepr (Am Chem Soc, Div Polym Chem) 38(1):6
21. Matyjaszewski K (1995) J Phys Org Chem 8:197
22. Matyjaszewski K, Gaynor S, Greszta D, Mardare D, Shigemoto T (1995) J Phys Org Chem 8:306
23. Matyjaszewski K (1993) Macromolecules 26:1787
24. Matyjaszewski K (1993) J Polym Sci, Part A: Polym Chem 31:995
25. Korolev GV, Marchenko AP (2000) Russ Chem Rev 69:409
26. Colombani D (1997) Prog Polym Sci 22:1649
27. Matyjaszewski K (1998) ACS Symp Ser 685:2

28. Sawamoto M, Kamigaito M (1999) In: Schlueter D (ed) Synthesis of polymers. VCH, Weinheim
29. Ajayaghosh A, Francis R (1998) Macromolecules 31:1436
30. Ajayaghosh A, Francis R (1999) J Am Chem Soc 121:6599
31. Borsig E, Lazar M, Capla M (1967) Makromol Chem 105:212
32. Braun D, Arcache G (1971) Makromol Chem 148:119
33. Qin S-H, Qiu K-Y, Swift G, Westmoreland DG, Wu S (1999) J Polym Sci, Part A: Polym Chem 37:4610
34. Sebenik A (1998) Prog Polym Sci 23:875
35. Matyjaszewski K (1998) (ed) Controlled radical polymerization. ACS Symp Ser 685, American Chemical Society, Washington, D.C.
36. Matyjaszewski K (ed) (2000) Controlled/living radical polymerization. ACS Symp Ser 768, American Chemical Society, Washington, D.C.
37. Solomon D, Rizzardo E, Cacioli P (1986) US Patent US 4,581,429
38. Georges MK, Veregin RPN, Kazmaier PM, Hamer GK (1993) Macromolecules 26:2987
39. Steenbock M, Klapper M, Muellen K, Bauer C, Hubrich M (1998) Macromolecules 31:5223
40. Wayland BB, Basickes L, Mukerjee S, Wei M, Fryd M (1997) Macromolecules 30:8109
41. Wayland BB, Mukerjee S, Poszmik G, Woska DC, Basickes L, Gridnev AA, Fryd M, Ittel SD (1998) ACS Symp Ser 685:305
42. Wang J-S, Matyjaszewski K (1995) J Am Chem Soc 117:5614
43. Kato M, Kamigaito M, Sawamoto M, Higashimura T (1995) Macromolecules 28:1721
44. Matyjaszewski K (1997) J Macromol Sci, Pure Appl Chem A34:1785
45. Patten TE, Matyjaszewski K (1998) Adv Mater 10(12):901
46. Patten TE, Matyjaszewski K (1999) Acc Chem Res 32:895
47. Matyjaszewski K (1999) Chem Eur J 5:3095
48. Sawamoto M, Kamigaito M (1996) Trends Polym Sci 4:371
49. Matyjaszewski K, Xia J (2001) Chem Rev 101:2921
50. Gaynor SG, Wang J-S, Matyjaszewski K (1995) Macromolecules 28:8051
51. Matyjaszewski K, Gaynor SG, Wang J-S (1995) Macromolecules 28:2093
52. Chiefari J, Chong YK, Ercole F, Krstina J, Jeffery J, Le TPT, Mayadunne RTA, Meijs GF, Moad CL, Moad G, Rizzardo E, Thang SH (1998) Macromolecules 31:5559
53. Chong YK, Le TPT, Moad G, Rizzardo E, Thang SH (1999) Macromolecules 32:2071
54. Mayadunne RTA, Rizzardo E, Chiefari J, Chong YK, Moad G, Thang SH (1999) Macromolecules 32:6977
55. Destarac M, Charmot D, Franck X, Zard SZ (2000) Macromol Rapid Comm 21:1035
56. Fischer H (1997) Macromolecules 30:5666
57. Rizzardo E (1987) Chem Aust 54:32
58. Rozantsev EG, Sholle VD (1971) Synthesis 190
59. Rozantsev EG, Sholle VD (1971) Synthesis 401
60. Moad G, Rizzardo E, Solomon DH, Beckwith ALJ (1992) Polym Bull 29:647
61. Hawker CJ (1994) J Am Chem Soc 116:11,185
62. Hawker CJ, Hedrick JL (1995) Macromolecules 28:2993
63. Catala J-M, Bubel F, Oulad Hammouch S (1995) Macromolecules 28:8441
64. Wang D, Wu Z (1998) Macromolecules 31:6727
65. Dao J, Benoit D, Hawker CJ (1998) J Polym Sci, Part A, Polym Chem 36:2161
66. Tirrell DA (1998) J Polym Sci, Part A, Polym Chem 36:2667
67. Matyjaszewski K, Woodworth BE, Zhang X, Gaynor SG, Metzner Z (1998) Macromolecules 31:5955
68. Miura Y, Hirota K, Moto H, Yamada B (1998) Macromolecules 31:4659
69. Zink M-O, Kramer A, Nesvadba P (2000) Macromolecules 33:7378
70. Benoit D, Grimaldi S, Finet J-P, Tordo P, Fontanille M, Gnanou Y (1997) In: Matyjaszewski K (ed) Controlled radical polymerization (ACS Symp Ser), vol 685. American Chemical Society, Washington, D.C., p 225

References

71. Benoit D, Chaplinski V, Braslau R, Hawker CJ (1999) J Am Chem Soc 121:3904
72. Grimaldi S, Finet J-P, Le Moigne F, Zeghdaoui A, Tordo P, Benoit D, Fountanille M, Gnanou Y (2000) Macromolecules 33:1141
73. Benoit D, Grimaldi S, Robin S, Finet J-P, Tordo P, Gnanou Y (2000) J Am Chem Soc 122:5929
74. Kharasch MS, Jensen EV, Urry WH (1945) Science 102:128
75. Minisci F (1975) Acc Chem Res 8:165
76. Asscher M, Vofsi D (1963) J Chem Soc 1963:3921
77. Curran DP (1988) Synthesis 489
78. Iqbal J, Bhatia B, Nayyar NK (1994) Chem Rev 94:519
79. Matyjaszewski K, Wang JS (1995) WO 9630421 US 5,763,548
80. Wang J-S, Matyjaszewski K (1995) Macromolecules 28:7901
81. Wang J-S, Matyjaszewski K (1995) Macromolecules 28:7572
82. Percec V, Barboiu B (1995) Macromolecules 28:7970
83. Boutevin B (2000) J Polym Sci, Part A: Polym Chem 38:3235
84. Lansalot M, Farcet C, Charleux B, Vairon J-P, Pirri R (1999) Macromolecules 32:7354
85. Mayadunne RTA, Rizzardo E, Chiefari J, Krstina J, Moad G, Postma A, Thang SA (2000) Macromolecules 33:243
86. Matyjaszewski K, Coessens V, Nakagawa Y, Xia J, Qiu J, Gaynor S, Coca S, Jasieczek C (1998) ACS Symp Ser 704:16
87. Coessens V, Pintauer T, Matyjaszewski K (2001) Prog Polym Sci 26:337
88. Coessens V, Matyjaszewski K (1999) Macromol Rapid Commun 20:127
89. Coessens V, Pyun J, Miller PJ, Gaynor SG, Matyjaszewski K (2000) Macromol Rapid Commun 21:103
90. Moad G, Chiefari J, Chong YK, Krstina J, Mayadunne RTA, Postma A, Rizzardo E, Thang SH (2000) Polym Int 49:993
91. Shipp DA, Wang J-L, Matyjaszewski K (1998) Macromolecules 31:8005
92. Ziegler MJ, Matyjaszewski K (2001) Macromolecules 34:415
93. de la Fuente JL, Fernandez-Garcia M, Fernandez-Sanz M, Madruga EL (2001) Macromolecules 34:5833
94. Matyjaszewski K, Shipp DA, McMurtry GP, Gaynor SG, Pakula T (2000) J Polym Sci, Part A: Polym Chem 38:2023
95. Hawker CJ, Elce E, Dao J, Volksen W, Russell TP, Barclay GG (1996) Macromolecules 29:2686
96. Fukuda T, Terauchi T, Goto A, Tsujii Y, Miyamoto T, Shimizu Y (1996) Macromolecules 29:3050
97. Pozzo J-L, Bouas-Laurent H, Deffieux A, Seidler D, Durr H (1997) Mol Cryst Liq Cryst Sci Technol, Sect A 298:437
98. Barclay GG, King M, Orellana A, Malenfant PRL, Sinta R, Malmstrom E, Ito H, Hawker CJ (1998) ACS Symp Ser 706:144
99. Yoshida E (1996) J Polym Sci, Part A: Polym Chem 34:2937
100. Yoshida E, Takiguchi Y (1999) Polym J 31(5):429
101. Baumert M, Mülhaupt R (1997) Macromol Rapid Commun 18:787
102. Schmidt-Naake G, Butz S (1996) Macromol Rapid Commun 17:661
103. Lokaj J, Vlcek P, Kriz J (1999) J Appl. Polym Sci 74:2378
104. Benoit D, Hawker CJ, Huang EE, Lin Z, Russell TP (2000) Macromolecules 33:1505
105. Baethge H, Butz S, Han C-H, Schmidt-Naake G (1999) Angew Makromol Chem 267:52
106. Mardare D, Matyjaszewski K (1994) Polym Prepr (Am Chem Soc, Div Polym Chem) 35(1):778
107. Devonport W, Michalak L, Malmstroem E, Mate M, Kurdi B, Hawker CJ, Barclay GG, Sinta R (1997) Macromolecules 30:1929
108. Greszta D, Matyjaszewski K (1997) J Polym Sci, Part A: Polym Chem 35:1857
109. Butz S, Baethge H, Schmidt-Naake G (1997) Macromol Rapid Commun 18:1049
110. Baethge H, Butz S, Schmidt-Naake G (1997) Macromol Rapid Commun 18:911

111. Fukuda T, Terauchi T, Goto A, Ohno K, Kobatake S, Yamada B (1996) Macromolecules 29:6393
112. Bignozzi MC, Ober CK, Novembre AJ, Knurek C (1999) Polym Bull 43(1):93
113. Jones RG, Yoon S, Nagasaki Y (1999) Polymer 40:2411
114. Wang J-S, Greszta D, Matyjaszewski K (1995) Polym Mater Sci Eng 73(2):416
115. Greszta D, Matyjaszewski K, Pakula T (1997) Polym Prepr (Am Chem Soc, Div Polym Chem) 38:709
116. Matyjaszewski K, Greszta D, Pakula T (1997) Polym Prepr (Am Chem Soc, Div Polym Chem) 38:707
117. Haddleton DM, Crossman MC, Hunt KH, Topping C, Waterson C, Suddaby KG (1997) Macromolecules 30:3992
118. Arehart SV, Matyjaszewski K (1999) Macromolecules 32:2221
119. Chambard G, Klumperman B (2000) ACS Symp Ser 768:197
120. Cassebras M, Pascual S, Polton A, Tardi M, Vairon J-P (1999) Macromol Rapid Commun 20:261
121. Matyjaszewski K, Shipp DA, Qiu J, Gaynor SG (2000) Macromolecules 33:2296
122. Matyjaszewski K, Qiu J, Shipp DA, Gaynor SG (2000) Macromol Symp 155:15
123. McQuillan BW, Paguio S (2000) Fusion Technol 38:108
124. Gao B, Chen X, Ivan B, Kops J, Batsberg W (1997) Polym Bull 39(5):559
125. Doerffler E, Patten TE (2000) Macromolecules 33:8911
126. Kotani Y, Kamigaito M, Sawamoto M (1998) Macromolecules 31:5582
127. Uegaki H, Kotani Y, Kamigato M, Sawamoto M (1998) Macromolecules 31:6756
128. Moineau G, Granel C, Dubois P, Jerome R, Teyssie P (1998) Macromolecules 31:542
129. Simal F, Demonceau A, Noels AF (1999) Angew Chem, Int Ed 38:538
130. Gaynor SG, Matyjaszewski K (1998) ACS Symp. Series 685:396
131. Coca S, Matyjaszewski K (1996) Polym Prepr (Am Chem Soc, Div Polym Chem) 37(1):573
132. Chen Q-C, Wu Z-Q, Wu J-R, Li Z-C, Li F-M (2000) Macromolecules 33:232
133. Tatemoto M, Oka M (1984) Contemp Topics Polym Sci 4:763
134. De Brouwer H, Schellekens MAJ, Klumperman B, Monteiro MJ, German AL (2000) J Polym Sci, Part A, Polym Chem 38:3596
135. Hirai H (1976) J Polym Sci, Macromol Rev 11:47
136. Hirai H, Tanabe T, Koinuma H (1979) J Polym Sci, Polym Chem Ed 17:843
137. Afchar-Momtaz J, Polton A, Tardi M, Sigwalt P (1985) Eur. Polym J 21:1067
138. Kirci B, Lutz JF, Matyjaszewski K (2002) Macromolecules 35:2448
139. Shinoda H, Matyjaszewski K (2001) Macromol Rapid Comm 22:1176
140. Bertin D, Boutevin B (1996) Polym Bull 37(3):337
141. Yoshida E, Fuji T (1997) J Polym Sci, Part A: Polym Chem 35:2371
142. Lacroix-Desmazes P, Delair T, Pichot C, Boutevin B (2000) J Polym Sci, Part A, Polym Chem 38:3845
143. Jousset S, Hammouch SO, Catala JM (1997) Macromolecules 30:6685
144. Oulad Hammouch S, Catala J-M (1996) Macromol Rapid Commun 17:149
145. Ohno K, Ejaz M, Fukuda T, Miyamoto T, Shimizu Y (1998) Macromol Chem Phys 199:291
146. Mariani M, Lelli M, Sparnacci K, Laus M (1999) J Polym Sci, Part A, Polym Chem 37:1237
147. Wan X, Tu Y, Zhang D, Zhou Q (1998) Chin. J Polym Sci 16:377
148. Wan X, Tu Y, Zhang D, Zhou Q (2000) Polym Int 49:243
149. Bignozzi MC, Ober CK, Laus M (1999) Macromol Rapid Commun 20:622
150. Gabaston LI, Furlong SA, Jackson RA, Armes SP (1999) Polymer 40:4505
151. Nowakowska M, Zapotoczny S, Karewicz A (2000) Macromolecules 33:7345
152. Listigovers NA, Georges MK, Odell PG, Keoshkerian B (1996) Macromolecules 29:8992
153. Zaremski MY, Stoyachenko YI, Hrenov VA, Garina ES, Lachinov MB, Golubev VB (1999) Russ Polym News 4:17

154. Barbosa CA, Gomes AS (1998) Polym Bull 41:15
155. Benoit D, Grimaldi S, Finet JP, Tordo P, Fontanille M, Gnanou Y (1997) Polym Prepr (Am Chem Soc, Div Polym Chem) 38:729
156. Grimaldi S, Finet J-P, Zeghdaoui A, Tordo P, Benoit D, Gnanou Y, Fontanille M, Nicol P, Pierson J-F (1997) Polym Prepr (Am Chem Soc, Div Polym Chem) 38:651
157. Georges MK, Hamer GK, Listigovers NA (1998) Macromolecules 31:9087
158. Keoshkerian B, Georges M, Quinlan M, Veregin R, Goodbrand B (1998) Macromolecules 31:7559
159. Benoit D, Harth E, Fox P, Waymouth RM, Hawker CJ (2000) Macromolecules 33:363
160. Bohrisch J, Wendler U, Jaeger W (1997) Macromol Rapid Commun 18:975
161. Wendler U, Bohrisch J, Jaeger W, Rother G, Dautzenberg H (1998) Macromol Rapid Commun 19:185
162. Li D, Brittain WJ (1998) Macromolecules 31:3852
163. Steenbock M, Klapper M, Muellen K, Pinhal N, Hubrich M (1996) Acta Polym 47:276
164. Lokaj J, Vlcek P, Kriz J (1997) Macromolecules 30:7644
165. Burguiere C, Dourges M-A, Charleux B, Vairon J-P (1999) Macromolecules 32:3883
166. Granel C, Dubois P, Jerome R, Teyssie P (1996) Macromolecules 29:8576
167. Kotani Y, Kato M, Kamigaito M, Sawamoto M (1996) Macromolecules 29:6979
168. Zhang X, Matyjaszewski K (1999) Macromolecules 32:1763
169. Beers KL, Boo S, Gaynor SG, Matyjaszewski K (1999) Macromolecules 32:5772
170. Xia J, Johnson T, Gaynor SG, Matyjaszewski K, DeSimone J (1999) Macromolecules 32:4802
171. Xia J, Zhang X, Matyjaszewski K (2000) ACS Symp Ser 760:207
172. Xia J, Gaynor SG, Matyjaszewski K (1998) Macromolecules 31:5958
173. Xia J, Matyjaszewski K (1997) Macromolecules 30:7697
174. Gobelt B, Matyjaszewski K (2000) Macromol Chem Phys 201:1619
175. Matyjaszewski K, Shipp DA, Wang J-L, Grimaud T, Patten TE (1998) Macromolecules 31:6836
176. Matyjaszewski K, Wang J-L, Grimaud T, Shipp D (1998) Macromolecules 31:1527
177. Moineau C, Minet M, Teyssie P, Jerome R (1999) Macromolecules 32:8277
178. Leclere P, Moineau G, Minet M, Dubois P, Jerome R, Bredas JL, Lazzaroni R (1999) Langmuir 15:3915
179. Moineau G, Minet M, Teyssie P, Jerome R (2000) Macromol Chem Phys 201:1108
180. Garcia MF, de la Fuente JL, Fernandez-Sanz M, Madruga EL (2001) Polymer 42:9405
181. Pyun J, Matyjaszewski K (2001) Chem Mater. 13:3436
182. Pyun J, Matyjaszewski K (2000) Macromolecules 33:217
183. Matyjaszewski K, Miller PJ, Kickelbick G, Nakagawa Y, Diamanti S, Pacis C (2000) ACS Symp. Series 729:270
184. Zou Y-S, Qiu Z-P, Zhuang R-C, Lin D-H, Dai L-Z (1998) Hecheng Huaxue 6:1
185. Zou Y, Zhuang R, Qiu Z, Dai L (1998) Chin. J React Polym 7:75
186. Wang X-S, Luo N, Ying S-K (1999) Polymer 40:4157
187. Liu Y, Wang L, Pan C (1999) Macromolecules 32:8301
188. Schubert U, Hochwimmer G, Spindler CE, Nuyken O (1999) Polym Bull 43:319
189. a) Xia J, Matyjaszewski K (1997) Macromolecules 30:7692, b) Wang J-S, Matyjaszewski K (1995) Macromolecules 28:7572
190. Xia J, Matyjaszewski K (1999) Macromolecules 32:5199
191. Qin D-Q, Qin S-H, Qiu K-Y (2000) Macromolecules 33:6987
192. Matyjaszewski K, Gaynor SG, Qiu J, Beers K, Coca S, Davis K, Muhlebach A, Xia J, Zhang X (2000) ACS Symp Ser 765:52
193. Qiu J, Charleux B, Matyjaszewski K (2001) Polimery (Warsaw, Poland) 46:663
194. Davis KA, Charleux B, Matyjaszewski K (2000) J Polym Sci Part A, Polym Chem 38:2274
195. Burguiere C, Pascual S, Bui C, Vairon J-P, Charleux B, Davis KA, Matyjaszewski K, Betremieux I (2001) Macromolecules 34:4439

196. Davis KA, Matyjaszewski K (2000) Macromolecules 33:4039
197. Davis KA, Matyjaszewski K (2001) Macromolecules 34:2101
198. Xia J, Zhang X, Matyjaszewski K (1999) Macromolecules 32:3531
199. Muehlebach A, Gaynor SG, Matyjaszewski K (1998) Macromolecules 31:6046
200. Teodorescu M, Matyjaszewski K (1999) Macromolecules 32:4826
201. Senoo M, Kotani Y, Kamigaito M, Sawamoto M (1999) Macromolecules 32:8005
202. Ohno K, Tsujii Y, Miyamoto T, Fukuda T, Goto M, Kobayashi K, Akaike T (1998) Macromolecules 31:1064
203. Ohno K, Tsujii Y, Fukuda T (1998) J Polym Sci, Part A: Polym Chem 36:2473
204. Haddleton DM, Ohno K (2000) Biomacromolecules 1:152
205. Marsh A, Khan A, Garcia M, Haddleton DM (2000) Chem Commun 2083
206. Wang XS, Jackson RA, Armes SP (2000) Macromolecules 33:255
207. Zhang Z, Ying S, Shi Z (1999) Polymer 40:5439
208. Coca S, Jasieczek CB, Beers KL, Matyjaszewski K (1998) J Polym Sci, A., Polym Chem 36:1417
209. Qiu J, Gaynor S, Matyjaszewski K (1999) Macromolecules 32:2872
210. Gaynor S, Qiu J, Matyjaszewski K (1998) Macromolecules 31:5951
211. Matyjaszewski K, Qiu J, Tsarevsky NV, Charleux B (2000) J Polym Sci, Part A: Polym Chem 38:4724
212. Qiu J, Charleux B, Matyjaszewski K (2001) Prog Polym Sci 26:2083
213. Haddleton DM, Jackson SG, Bon SAF (2000) J Am Chem Soc 122:1542
214. Ashford EJ, Naldi V, O'Dell R, Billingham NC, Armes SP (1999) Chem Commun 1285
215. Farcet C, Lansalot M, Pirri R, Vairon JP, Charleux B (2000) Macromol Rapid Commun 21:921
216. Matyjaszewski K, Coca S, Gaynor SG, Wei M, Woodworth BE (1998) Macromolecules 31:5967
217. Matyjaszewski K, Coca S, Gaynor SG, Wei M, Woodworth BE (1997) Macromolecules 30:7348
218. Matyjaszewski K, Beers KL, Woodworth B, Metzner Z (2001) J Chem Ed 78:547
219. Beers KL, Woodworth B, Matyjaszewski K (2001) J Chem Ed 78:544
220. Zhang X, Matyjaszewski K (1999) Macromolecules 32:7349
221. Xia J, Paik H-j, Matyjaszewski K (1999) Macromolecules 32:8310
222. Coca S, Matyjaszewski K (1997) Macromolecules 30:2808
223. Gaynor SG, Matyjaszewski K (1997) Macromolecules 30:4241
224. Jankova K, Chen X, Kops J, Batsberg W (1998) Macromolecules 31:538
225. Chen X, Gao B, Kops J, Batsberg W (1998) Polymer 39:911
226. Jankova K, Truelsen JH, Chen X, Kops J, Batsberg W (1999) Polym Bull 42:153
227. Cheng S, Xu Z, Yuan J, Ji P, Xu J, Ye M, Shi L (2000) J Appl Polym Chem 77:2882
228. Bednarek M, Biedron T, Kubisa P (1999) Macromol Rapid Commun 20:59
229. Bednarek M, Biedron T, Kubisa P (2000) Macromol Chem Phys 201:58
230. Wang X-S, Luo N, Ying S-K, Liu Q (2000) Eur Polym J 36:149
231. Yoshida E, Nakamura M (1998) Polym J 30(11):915
232. Matyjaszewski K, Miller PJ, Fossum E, Nakagawa Y (1998) Appl Organomet Chem 12:667
233. Nakagawa Y, Miller PJ, Matyjaszewski K (1998) Polymer 39:5163
234. Miller PJ, Matyjaszewski K (1999) Macromolecules 32:8760
235. Jankova K, Kops J, Chen X, Batsberg W (1999) Macromol Rapid Commun 20:219
236. Schellekens MAJ, Klumperman B, van der Linde R (2001) Macromol Chem Phys 202:1595
237. Gaynor SG, Edelman SZ, Matyjaszewski K (1997) Polym Prep (Am Chem Soc, Polym Div) 38(1):703
238. Tong J-D, Ni S, Winnik MA (2000) Macromolecules 33:1482
239. Tong JD, Moineau G, Leclere P, Bredas JL, Lazzaroni R, Jerome R (2000) Macromolecules 33:470

240. Lusten L, Cordina GP-G, Jones RG, Schue F (1998) Eur Polym J 34:1829
241. Tsolakis PK, Koulouri EG, Kallitsis JK (1999) Macromolecules 32:9054
242. Matyjaszewski K, Coca S, Gaynor SG, Wei M, Woodworth BE (1998) Macromolecules 31:5967
243. Stalmach U, de Boer B, Videlot C, van Hutten PF, Hadziioannou G (2000) J Am Chem Soc 122:5464
244. Li IQ, Howell BA, Dineen MT, Kastl PE, Lyons JW, Meunier DM, Smith PB, Priddy DB (1997) Macromolecules 30:5195
245. Yoshida E, Tanimoto S (1997) Macromolecules 30:4018
246. Paik H-j, Teodorescu M, Xia J, Matyjaszewski K (1999) Macromolecules 32:7023
247. Destarac M, Pees B, Boutevin B (2000) Macromol Chem Phys 201:1189
248. Destarac M, Boutevin B (1999) Macromol Rapid Commun 20:641
249. Zhang Z, Ying S, Shi Z (1999) Polymer 40:1341
250. Destarac M, Matyjaszewski K, Silverman E, Ameduri B, Boutevin B (2000) Macromolecules 33:4613
251. Alkan S, Toppare L, Hepuzer Y, Yagci Y (1999) J Polym Sci, Part A, Polym Chem 37:4218
252. Leduc MR, Hawker CJ, Dao J, Frechet JMJ (1996) J Am Chem Soc 118:11,111
253. Matyjaszewski K, Shigemoto T, Frechet JMJ, Leduc M (1996) Macromolecules 29:4167
254. Leduc MR, Hayes W, Frechet JMJ (1998) J Polym Sci, Part A: Polym Chem 36:1
255. Emrick T, Hayes W, Frechet JMJ (1999) J Polym Sci, Part A: Polym Chem 37:3748
256. Hovestad NJ, van Koten G, Bon SAF, Haddleton DM (2000) Macromolecules 33:4048
257. Frechét JMJ, Henmi M, Gitsov I, Aoshima S, Leduc M, Grubbs RB (1995) Science 269:1080
258. Gaynor SG, Edelman S, Matyjaszewski K (1996) Macromolecules 29:1079
259. Matyjaszewski K, Gaynor SG, Kulfan A (1997) Macromolecules 30:5192
260. Matyjaszewski K, Gaynor SG, Mueller A (1997) Macromolecules 30:7034
261. Matyjaszewski K, Gaynor SG (1997) Macromolecules 30:7042
262. Matyjaszewski K, Pyun J, Gaynor SG (1998) Macromol Rapid Commun 19:665
263. Cheng G, Simon PFW, Hartenstein M, Muller AHE (2000) Macromol Rapid Commun 21:846
264. Zhang X, Chen Y, Gong A, Chen C, Xi F (1999) Polym Int 48:896
265. An S-G, Cho C-G (2000) Polym Prepr (Am Chem Soc, Div Polym Chem) 41(2):1671
266. Jiang X, Zhong Y, Yan D, Yu H, Zhang D (2000) J Appl Polym Sci 78:1992
267. Sunder A, Hanselmann R, Frey H, Mulhaupt R (1999) Macromolecules 32:4240
268. Maier S, Sunder A, Frey H, Mulhaupt R (2000) Macromol Rapid Commun 21:226
269. Coca S, Matyjaszewski K (1997) J Polym Sci, Part A, Polym Chem 35:3595
270. Ivan B, Chen X, Kops J, Batsberg W (1998) Macromol Rapid Commun 19:15
271. Yoshida E, Sugita A (1996) Macromolecules 29:6422
272. Yoshida E, Sugita A (1998) J Polym Sci, Part A: Polym Chem 36:2059
273. Yagci Y, Duez AB, Oenen A (1997) Polymer 38:2861
274. Kajiwara A, Matyjaszewski K (1998) Macromolecules 31:3489
275. Liu Y, Wan X, Ying S (1998) Hecheng Xiangjiao Gongye 21:306
276. Liu Y, Wan X, Ying S (1998) Hecheng Xiangjiao Gongye 21:305
277. Liu Y, Ying S, Wan X (1999) Polym Prepr (Am Chem Soc, Div Polym Chem) 40:1053
278. Coessens V, Matyjaszewski K (1999) J Macromol Sci, Pure Appl Chem A36:667
279. Xu Y, Pan C (2000) J Polym Sci, Part A, Polym Chem 38:337
280. Xu Y, Pan C, Tao L (2000) J Polym Sci, Part A, Polym Chem 38:436
281. Yildirim TG, Hepuzer Y, Hizal G, Yagci Y (1999) Polymer 40:3885
282. Kobatake S, Harwood HJ, Quirk RP, Priddy DB (1998) Macromolecules 31:3735
283. Miura Y, Hirota K, Moto H, Yamada B (1999) Macromolecules 32:8356
284. Acar MH, Matyjaszewski K (1999) Macromol Chem Phys 200:1094
285. Liu B, Liu F, Luo N, Ying S (1998) Hecheng Xiangjiao Gongye 21:304
286. Liu F, Liu B, Luo N, Ying S (1998) Hecheng Xiangjiao Gongye 21:303
287. Liu B, Liu F, Luo N, Ying S-K, Liu Q (2000) Chin J Polym Sci 18:39

288. Liu F, Liu B, Luo N, Ying S, Liu Q (2000) Chem Res Chin Univ 16:72
289. Liu F, Ying S, Luo N, Liu B (1999) Polym Prepr (Am Chem Soc, Div Polym Chem) 40:1032
290. Yoshida E, Osagawa Y (1998) Macromolecules 31:1446
291. Wang Y, Chen S, Huang J (1999) Macromolecules 32:2480
292. Coca S, Paik H-j, Matyjaszewski K (1997) Macromolecules 30:6513
293. a) Bielawski CW, Morita T, Grubbs RH (2000) Macromolecules 33:678, b) Bielawski CW, Louie J, Grubbs RH (2000) J. Am. Chem. Soc. 122:12872
294. Hawker CJ, Frechet JMJ, Grubbs RB, Dao J (1995) J Am Chem Soc 117:10,763
295. Stehling UM, Malmstroem EE, Waymouth RM, Hawker CJ (1998) Macromolecules 31:4396
296. Miwa Y, Yamamoto K, Sakaguchi M, Shimada S (1999) Macromolecules 32:8234
297. Paik H-j, Gaynor SG, Matyjaszewski K (1998) Macromol Rapid Commun 19:47
298. Percec V, Asgarzadeh F (2001) J Polym Sci, Part A: Polym Chem 39:1120
299. Liu S, Sen A (2000) Macromolecules 33:5106
300. Fonagy T, Ivan B, Szesztay M (1998) Macromol Rapid Commun 19:479
301. Hong SC, Pakula T, Matyjaszewski K (2001) Macromol Chem Phys 202(17):3392–3402
302. Pan Q, Liu S, Xie J, Jiang M (1999) J Polym Sci, Part A, Polym Chem 37:2699
303. Matyjaszewski K, Teodorescu M, Miller PJ, Peterson ML (2000) J Polym Sci, Part A: Polym Chem 38:2440
304. Hong SC, Matyjaszewski K, Gottfried A, Brookhart M (2001) Polym Mat Sci Eng 85:363
305. Grubbs RB, Hawker CJ, Dao J, Frechet JMJ (1997) Angew Chem, Int Ed Engl 36:270
306. Liu B, Hu C, Hua F, Yang Y, He J (1999) Hecheng Xiangjiao Gongye 22:373
307. Beers KL, Gaynor SG, Matyjaszewski K, Sheiko SS, Moeller M (1998) Macromolecules 31:9413
308. Boerner HG, Beers K, Matyjaszewski K, Sheiko SS, Moeller M (2001) Macromolecules 34:4375
309. Yamada K, Miyazaki M, Ohno K, Fukuda T, Minoda M (1999) Macromolecules 32:290
310. a) Beers KL, Gaynor SG, Matyjaszewski K, Sheiko SS, Prokhorova SA, Moller M (1999) Polym Prepr (Am Chem Soc, Div Polym Chem) 40:446, b) Boerner HG, Duran D, Matyjaszewski K, da Silva M, Sheiko SS (2002) Macromolecules 35:3387
311. Sheiko SS, Prokhorova SA, Beers KL, Matyjaszewski K, Potemkin II, Khokhlov AR, Moeller M (2001) Macromolecules 34:8354
312. Pakula T, Minkin P, Beers KL, Matyjaszewski K (2001) Polym Mat Sci Eng 84:1006
313. Cheng G, Boeker A, Zhang M, Krausch G, Mueller AHE (2001) Macromolecules 34:6883
314. Truelsen JH, Kops J, Batsberg W (2000) Macromol Rapid Commun 21:98
315. Hawker CJ, Mecerreyes D, Elce E, Dao J, Hedrick JL, Barakat I, Dubois P, Jerome R, Volksen I (1997) Macromol Chem Phys 198:155
316. Wang Y, Huang J (1998) Macromolecules 31:4057
317. Matyjaszewski K, Beers KL, Kern A, Gaynor SG (1998) J Polym Sci, Part A, Polym Chem 36:823
318. Roos SG, Mueller AHE, Matyjaszewski K (1999) Macromolecules 32:8331
319. Roos SG, Muller AHE, Matyjaszewski K (2000) ACS Symp Ser 768:361
320. Shinoda H, Miller PJ, Matyjaszewski K (2001) Macromolecules 34:3186
321. Shinoda H, Matyjaszewski K (2001) Macromolecules 34:6243
322. Moad G (1999) Prog Polym Sci 24:81
323. Desai A, Atkinson N, Rivera F Jr, Devonport W, Rees I, Branz SE, Hawker CJ (2000) J Polym Sci, Part A, Polym Chem 38:1033
324. Huang X, Doneski LJ, Wirth MJ (1998) Anal Chem 70:4023
325. Huang X, Wirth MJ (1999) Macromolecules 32:1694
326. Ejaz M, Yamamoto S, Ohno K, Tsujii Y, Fukuda T (1998) Macromolecules 31:5934
327. Husemann M, Malmstroem EE, McNamara M, Mate M, Mecerreyes D, Benoit DG, Hedrick JL, Mansky P, Huang E, Russell TP, Hawker CJ (1999) Macromolecules 32:1424

328. Matyjaszewski K, Miller PJ, Shukla N, Immaraporn B, Gelman A, Luokala BB, Siclovan TM, Kickelbick G, Vallant T, Hoffman H, Pakula T (1999) Macromolecules 32:8716
329. a) Zhao B, Brittain WJ (1999) J Am Chem Soc 121:3557, b) Zhao B, Brittian WJ (2000) Prog. Pol. Sci. 25:677
330. Zhao B, Brittain WJ, Zhou W, Cheng SZD (2000) J Am Chem Soc 122:2407
331. Sedjo RA, Mirous BK, Brittain WJ (2000) Macromolecules 33:1492
332. Shah RR, Merreceyes D, Husemann M, Rees I, Abbott NL, Hawker CJ, Hedrick JL (2000) Macromolecules 33:597
333. Kim J-B, Bruening ML, Baker GL (2000) J Am Chem Soc 122:7616
334. Weimer MW, Chen H, Giannelis EP, Sogah DY (1999) J Am Chem Soc 121:1615
335. Bottcher H, Hallensleben ML, Nuss S, Wurm H (2000) Polym Bull 44(2):223
336. von Werne T, Patten TE (1999) J Am Chem Soc 121:7409
337. von Werne T, Patten TE (2001) J Am Chem Soc 123:7497
338. Pyun J, Matyjaszewski K, Kowalewski T, Savin D, Patterson G, Kickelbick G, Huesing N (2001) J Am Chem Soc 123:9445
339. Pyun J, Matyjaszewski K, Kowalewski T, Savin D, Patterson G, Kickelbick G, Huesing N (2001) Polym Prepr (Am Chem Soc, Div Polym Chem) 42:223
340. Farmer SC, Patten TE (2001) Polym Prepr (Am Chem Soc, Div Polym Chem) 42:578
341. Angot S, Ayres N, Bon SAF, Haddleton DM (2001) Macromolecules 34:768
342. Huang X, Wirth MJ (1997) Anal Chem 69:4577
343. Mandal TK, Fleming MS, Walt DR (2000) Chem Mater 12:3481
344. Guerrini MM, Charleux B, Vairon J-P (2000) Macromol Rapid Commun 21:669
345. Matyjaszewski K, Miller PJ, Pyun J, Kickelbick G, Diamanti S (1999) Macromolecules 32:6526
346. Ueda J, Kamigaito M, Sawamoto M (1998) Macromolecules 31:6762
347. Ueda J, Matsuyama M, Kamigaito M, Sawamoto M (1998) Macromolecules 31:557
348. Angot S, Shanmugananda Murthy K, Taton D, Gnanou Y (1998) Macromolecules 31:7218
349. Heise A, Hedrick JL, Frank CW, Miller RD (1999) J Am Chem Soc 121:8647
350. Heise A, Hedrick JL, Trollsas M, Miller RD, Frank CW (1999) Macromolecules 31:231
351. Heise A, Nguyen C, Malek R, Hedrick JL, Frank CW, Miller RD (2000) Macromolecules 33:2346
352. Hedrick JL, Trollss M, Hawker CJ, Atthoff B, Claesson H, Heise A, Miller RD, Mecerreyes D, Jerome R, Dubois P (1998) Macromolecules 31:8691
353. Xu Y, Pan C (2000) Macromolecules 33:4750
354. Keszler B, Fenyvesi G, Kennedy JP (2000) J Polym Sci, Part A: Polym Chem 38:706
355. Angot S, Taton D, Gnanou Y (2000) Macromolecules 33:5418
356. Xia J, Zhang X, Matyjaszewski K (1999) Macromolecules 32:4482
357. Zhang X, Xia J, Matyjaszewski K (2000) Macromolecules 33:2340
358. Puts RD, Sogah DY (1997) Macromolecules 30:7050
359. Weimer MW, Scherman OA, Sogah DY (1998) Macromolecules 31:8425
360. Hawker CJ, Hedrick JL, Malmstrom E, Trollsas M, Stehling UM, Waymouth RM (1998) Solvent-free polymerization and processes (ACS Symp Ser), vol 713, p 127
361. Jerome R, Mecerreyes D, Tian D, Dubois P, Hawker CJ, Trollsas M, Hedrick JL (1998) Macromol Symp 132:385
362. Mecerreyes D, Moineau G, Dubois P, Jerome R, Hedrick JL, Hawker CJ, Malmstrom EE, Trollsas M (1998) Angew Chem, Int Ed 37:1274
363. Klaener G, Trollsas M, Heise A, Husemann M, Atthof B, Hawker CJ, Hedrick JL, Miller RD (1999) Macromolecules 32:8227
364. Mecerreyes D, Atthoff B, Boduch KA, Trollsaas M, Hedrick JL (1999) Macromolecules 32:5175
365. Hawker CJ, Hedrick JL, Malmstrom EE, Trollsas M, Mecerreyes D, Moineau G, Dubois P, Jerome R (1998) Macromolecules 31:213

366. Mecerreyes D, Humes J, Miller RD, Hedrick JL, Detrembleur C, Lecomte P, Jerome R, San Roman J (2000) Macromol Rapid Commun 21:779
367. Matyjaszewski K (1998) Macromolecules 31:4710
368. Hong SC, Paik HP, Matyjaszewski K (2001) Macromolecules 34:5099
369. Husemann M, Morrison M, Benoit D, Frommer J, Mate CM, Hinsberg WD, Hedrick JL, Hawker CJ (2000) J Am Chem Soc 122:1844
370. Matyjaszewski K (2001) Macromol Symp 174:51
371. Huan K, Bes L, Haddleton DM, Khoshdel E (2001) J Polym Sci, Part A: Polym Chem 39:1833
372. Hawker CJ (1995) Angew Chem Int Ed Engl 34:1456

Received: December 2001

List of Abbreviations

AEMI	N-(2-acetoxyethyl) maleimide
AFM	atomic force microscopy
AIBN	2,2'-diazobisisobutyralnitrile
AMBEP	2,2'-azobis[2-methyl-N-(2-(2-bromoisobutyryloxy)ethyl)-propionamide
AMCBP	2,2'-azobis[2-methyl-N-(2-(4-chloromethylbenzoyloxy)-(ethyl)propionamide
AN	acrylonitrile
AcOSt	p-acetoxystyrene
ATRA	atom transfer radical addition
ATRP	atom transfer radical polymerization
β-CD	β-cyclodextrin
BD	1,3-butadiene
BIEA	2-(2-bromoisobutyryloxy)ethyl acrylate
BPEA	2-(2-bromopropionyloxy)ethyl acrylate
BPEM	2-(2-bromopropionyloxy)ethyl methacrylate
BPO	benzoyl peroxide
BPPN	N-$tert$-butyl- [1-phenyl-(2-methylpropyl)] nitroxide
bpy	2,2'-bipyridine
BrSt	p-bromostyrene
BzMA	benzyl methacrylate
CHO	cyclohexene oxide
CL	ε-caprolactone
CMI	cyclohexylmaleimide
CMSt	chloromethylstyrene
CPD	cyclopentadiene
CRP	controlled radical polymerization
D_3	hexamethylcyclotrisiloxane

DBX	*p*-dibromoxylene
DCP	dicumyl peroxide
DEPN	*N-tert*-butyl-*N*-[1-diethylphosphono-(2,2-dimethylpropyl)]nitroxide
dHbpy	4,4'-diheptyl-2,2'-bipyridine
DHF	2,7-dibromo-9,9-dihexylfluorene
DLS	dynamic light scattering
DMA	*N,N*-dimethylacrylamide
DMAEMA	*N,N*-dimethylaminoethyl methacrylate
DMAMS	4-(dimethylamino)methylstyrene
dMbpy	4,4'-dimethyl-2,2'-bipyridine
DMF	*N,N*-dimethylformamide
dNbpy	4,4'-di-(5-nonyl)-2,2'-bipyridine
DOP	1,3-dioxepane
$dR_{f6}bpy$	4,4'-di(tridecafluoro-1,1,2,2,3,3-hexahydrononyl)-2,2'-bipyridine
EAD	ethylene adipate
EB	ethylene-*co*-butylene
EBPBB	4'-ethylbiphenyl-4-(4-propenoyloxy-butyloxy)benzoate
EGMAFO	ethylene glycol mono-methacrylate mono-perfluorooctanoate
EMA	ethyl methacrylate
EO	ethylene oxide
EPSt	epoxystyrene
FNEMA	2-[(per-fluorononenyl)oxy] ethyl methacrylate
FOA	1,1-dihydroperfluorooctyl acrylate
FOMA	1,1-dihydroperfluorooctyl methacrylate
GPC	gel permeation chromatography
HEA	2-hydroxyethyl acrylate
HPMA	*N*-(2-hydroxypropyl)-methacrylamide
HEMA	2-hydroxyethyl methacrylate
HTEMPO	4-hydroxy-TEMPO
IA	isobornyl acrylate
IB	isobutylene
IBVE	isobutyl vinyl ether
IP	isoprene
M_n	number average molecular weight
M_w/M_n	polydispersity
MA	methyl acrylate
MAA	methacrylic acid
MAGlc	3-*O*-methacryloyl-1,2:5,6-D-glucofuranose
MAh	maleic anhydride
MAIpGlc	3-*O*-methacryloyl-1,2:5,6-di-*O*-isopropylidene-D-glucofuranose
MA-POSS	(polyhedral silsesquioxane) methacrylate

MeSt	α-methylstyrene
Me$_4$cyclam	1,4,8,11-tetramethyl-1,4,8,11-tetraazacyclotetradecane
Me$_6$TREN	tris [2-(dimethylamino)ethyl] amine
MM	macromonomer
MMA	methyl methacrylate
MO	2-methyloxazoline
MPCS	2,5-bis [(4-methoxyphenyl)oxycarbonyl] styrene
MPSi	methylphenylsilylene
MPVB	[(4'-methoxyphenyl)4-oxybenzoate]-6-hexyl (4-vinylbenzoate)
MSSQ	methylsilsesquioxane
MOTEMPO	4-methoxy-TEMPO
MVB-TMS	trimethylsilyl methyl(4-vinylbenzoate)
NaVB	sodium 4-vinyl benzoate
NB	norbornene
nBA	n-butyl acrylate
nBMA	n-butyl methacrylate
NPMA	p-nitrophenyl methacrylate
NVC	N-vinyl carbazole
NVP	N-vinyl pyrrolidinone
OEGMA	oligo(ethylene oxide) methacrylate
OPMA	n-octyl-2-pyridylmethanimine
OTEMPO	4-oxy-TEMPO
pSt	polystyrene
pDMS	poly(dimethylsiloxane)
PECl	1-phenylethyl chloride
PEG	poly(ethylene glycol)
PG	polyglycerol
PIMS	phthalimide methylstyrene
PMDETA	N,N,N',N'',N''-pentamethyldiethylenetriamine
PMI	N-phenylmaleimide
PO	phenyl oxazoline
pP	polypropylene
PV	phenylenevinylene
PY	pyrrole
RAFT	reversible addition fragmentation chain transfer
rATRP	reverse atom transfer radical polymerization
ROMP	ring opening metathesis polymerization
ROP	ring opening polymerization
scCO$_2$	super critical carbon dioxide
SEP	styrene-b-ethylene-co-propylene
SO	styrene oxide
SSC	sodium 4-styrenecarboxylate

List of Abbreviations

SSt	sodium 4-styrenesulfonate
St	Styrene
tBA	*tert*-butyl acrylate
tBOSt	4-*tert*-butoxy styrene
tBuSt	4-*tert*-butyl styrene
TCAC	trichloroacetyl chloride
TEM	transmission electron microscopy
TEMPO	2,2,6,6-tetramethyl-1-piperidinyloxy
T_g	glass transition temperature
THF	tetrahydrofuran
TMPAH	2,2,5-trimethyl-3-(1-phenylethoxy)-4-phenyl-3-azahexane
TsCl	*p*-toluenesulfonyl chloride
VA	vinyl acetate
VBC	vinyl benzyl chloride (*p*-chloromethyl styrene)
VC	vinyl chloride
VDF	vinylidene fluoride
VN	vinyl naphthalene
VP	vinylpyridine
XPS	X-ray photoelectron spectroscopy

Author Index Volumes 101–159

Author Index Volumes 1–100 see Volume 100

de, Abajo, J. and *de la Campa, J.G.*: Processable Aromatic Polyimides. Vol. 140, pp. 23-60.
Adolf, D. B. see Ediger, M. D.: Vol. 116, pp. 73-110.
Aharoni, S. M. and *Edwards, S. F.*: Rigid Polymer Networks. Vol. 118, pp. 1-231.
Albertsson, A.-C., Varma, I. K.: Aliphatic Polyesters: Synthesis, Properties and Applications. Vol. 157, pp. 99–138.
Albertsson, A.-C. see Edlund, U.: Vol. 157, pp. 53-98.
Albertsson, A.-C. see Söderqvist Lindblad, M.: Vol. 157, pp. 139–161.
Albertsson, A.-C. see Stridsberg, K. M.: Vol. 157, pp. 27–51.
Améduri, B., Boutevin, B. and *Gramain, P.*: Synthesis of Block Copolymers by Radical Polymerization and Telomerization. Vol. 127, pp. 87-142.
Améduri, B. and *Boutevin, B.*: Synthesis and Properties of Fluorinated Telechelic Monodispersed Compounds. Vol. 102, pp. 133-170.
Amselem, S. see Domb, A. J.: Vol. 107, pp. 93-142.
Andrady, A. L.: Wavelenght Sensitivity in Polymer Photodegradation. Vol. 128, pp. 47-94.
Andreis, M. and *Koenig, J. L.*: Application of Nitrogen-15 NMR to Polymers. Vol. 124, pp. 191-238.
Angiolini, L. see Carlini, C.: Vol. 123, pp. 127-214.
Anseth, K. S., Newman, S. M. and *Bowman, C. N.*: Polymeric Dental Composites: Properties and Reaction Behavior of Multimethacrylate Dental Restorations. Vol. 122, pp. 177-218.
Antonietti, M. see Cölfen, H.: Vol. 150, pp. 67-187.
Armitage, B. A. see O'Brien, D. F.: Vol. 126, pp. 53-58.
Arndt, M. see Kaminski, W.: Vol. 127, pp. 143-187.
Arnold Jr., F. E. and *Arnold, F. E.*: Rigid-Rod Polymers and Molecular Composites. Vol. 117, pp. 257-296.
Arora, M. see Kumar, M.N.V.R.: Vol. 160, pp. 45-118.
Arshady, R.: Polymer Synthesis via Activated Esters: A New Dimension of Creativity in Macromolecular Chemistry. Vol. 111, pp. 1-42.

Bahar, I., Erman, B. and *Monnerie, L.*: Effect of Molecular Structure on Local Chain Dynamics: Analytical Approaches and Computational Methods. Vol. 116, pp. 145-206.
Ballauff, M. see Dingenouts, N.: Vol. 144, pp. 1-48.
Baltá-Calleja, F. J., González Arche, A., Ezquerra, T. A., Santa Cruz, C., Batallón, F., Frick, B. and *López Cabarcos, E.*: Structure and Properties of Ferroelectric Copolymers of Poly(vinylidene) Fluoride. Vol. 108, pp. 1-48.
Barnes, M. D. see Otaigbe, J.U.: Vol. 154, pp. 1-86.
Barshtein, G. R. and *Sabsai, O. Y.*: Compositions with Mineralorganic Fillers. Vol. 101, pp.1-28.
Baschnagel, J., Binder, K., Doruker, P., Gusev, A. A., Hahn, O., Kremer, K., Mattice, W. L., Müller-Plathe, F., Murat, M., Paul, W., Santos, S., Sutter, U. W., Tries, V.: Bridging the Gap Between Atomistic and Coarse-Grained Models of Polymers: Status and Perspectives. Vol. 152, pp. 41-156.
Batallán, F. see Baltá-Calleja, F. J.: Vol. 108, pp. 1-48.
Batog, A. E., Pet'ko, I. P., Penczek, P.: Aliphatic-Cycloaliphatic Epoxy Compounds and Polymers. Vol. 144, pp. 49-114.

Barton, J. see Hunkeler, D.: Vol. 112, pp. 115-134.
Bell, C. L. and *Peppas, N. A.*: Biomedical Membranes from Hydrogels and Interpolymer Complexes. Vol. 122, pp. 125-176.
Bellon-Maurel, A. see Calmon-Decriaud, A.: Vol. 135, pp. 207-226.
Bennett, D. E. see O'Brien, D. F.: Vol. 126, pp. 53-84.
Berry, G.C.: Static and Dynamic Light Scattering on Moderately Concentraded Solutions: Isotropic Solutions of Flexible and Rodlike Chains and Nematic Solutions of Rodlike Chains. Vol. 114, pp. 233-290.
Bershtein, V. A. and *Ryzhov, V. A.*: Far Infrared Spectroscopy of Polymers. Vol. 114, pp. 43-122.
Bigg, D. M.: Thermal Conductivity of Heterophase Polymer Compositions. Vol. 119, pp. 1-30.
Binder, K.: Phase Transitions in Polymer Blends and Block Copolymer Melts: Some Recent Developments. Vol. 112, pp. 115-134.
Binder, K.: Phase Transitions of Polymer Blends and Block Copolymer Melts in Thin Films. Vol. 138, pp. 1-90.
Binder, K. see Baschnagel, J.: Vol. 152, pp. 41-156.
Bird, R. B. see Curtiss, C. F.: Vol. 125, pp. 1-102.
Biswas, M. and *Mukherjee, A.*: Synthesis and Evaluation of Metal-Containing Polymers. Vol. 115, pp. 89-124.
Biswas, M. and *Sinha Ray, S.*: Recent Progress in Synthesis and Evaluation of Polymer-Montmorillonite Nanocomposites. Vol. 155, pp. 167-221.
Bolze, J. see Dingenouts, N.: Vol. 144, pp. 1-48.
Bosshard, C.: see Gubler, U.: Vol. 158, pp. 123-190.
Boutevin, B. and *Robin, J. J.*: Synthesis and Properties of Fluorinated Diols. Vol. 102. pp. 105-132.
Boutevin, B. see Amédouri, B.: Vol. 102, pp. 133-170.
Boutevin, B. see Améduri, B.: Vol. 127, pp. 87-142.
Bowman, C. N. see Anseth, K. S.: Vol. 122, pp. 177-218.
Boyd, R. H.: Prediction of Polymer Crystal Structures and Properties. Vol. 116, pp. 1-26.
Briber, R. M. see Hedrick, J. L.: Vol. 141, pp. 1-44.
Bronnikov, S. V., Vettegren, V. I. and *Frenkel, S. Y.*: Kinetics of Deformation and Relaxation in Highly Oriented Polymers. Vol. 125, pp. 103-146.
Brown, H. R. see Creton, C.: Vol. 156, pp. 53-135.
Bruza, K. J. see Kirchhoff, R. A.: Vol. 117, pp. 1-66.
Budkowski, A.: Interfacial Phenomena in Thin Polymer Films: Phase Coexistence and Segregation. Vol. 148, pp. 1-112.
Burban, J. H. see Cussler, E. L.: Vol. 110, pp. 67-80.
Burchard, W.: Solution Properties of Branched Macromolecules. Vol. 143, pp. 113-194.

Calmon-Decriaud, A. Bellon-Maurel, V., Silvestre, F.: Standard Methods for Testing the Aerobic Biodegradation of Polymeric Materials. Vol 135, pp. 207-226.
Cameron, N. R. and *Sherrington, D. C.*: High Internal Phase Emulsions (HIPEs)-Structure, Properties and Use in Polymer Preparation. Vol. 126, pp. 163-214.
de la Campa, J. G. see de Abajo, , J.: Vol. 140, pp. 23-60.
Candau, F. see Hunkeler, D.: Vol. 112, pp. 115-134.
Canelas, D. A. and *DeSimone, J. M.*: Polymerizations in Liquid and Supercritical Carbon Dioxide. Vol. 133, pp. 103-140.
Canva, M., Stegeman, G. I.: Quadratic Parametric Interactions in Organic Waveguides. Vol. 158, pp. 87-121.
Capek, I.: Kinetics of the Free-Radical Emulsion Polymerization of Vinyl Chloride. Vol. 120, pp. 135-206.
Capek, I.: Radical Polymerization of Polyoxyethylene Macromonomers in Disperse Systems. Vol. 145, pp. 1-56.
Capek, I.: Radical Polymerization of Polyoxyethylene Macromonomers in Disperse Systems. Vol. 146, pp. 1-56.

Capek, I. and *Chern, C.-S.*: Radical Polymerization in Direct Mini-Emulsion Systems. Vol. 155, pp. 101-166.
Carlesso, G. see Prokop, A.: Vol. 160, pp. 119-174.
Carlini, C. and *Angiolini, L.*: Polymers as Free Radical Photoinitiators. Vol. 123, pp. 127-214.
Carter, K. R. see Hedrick, J. L.: Vol. 141, pp. 1-44.
Casas-Vazquez, J. see Jou, D.: Vol. 120, pp. 207-266.
Chandrasekhar, V.: Polymer Solid Electrolytes: Synthesis and Structure. Vol 135, pp. 139-206
Chang, J.Y. see Han, M. J.: Vol. 153, pp. 1-36.
Charleux, B., Faust R.: Synthesis of Branched Polymers by Cationic Polymerization. Vol. 142, pp. 1-70.
Chen, P. see Jaffe, M.: Vol. 117, pp. 297-328.
Chern, C.-S. see Capek, I.: Vol. 155, pp. 101-166.
Choe, E.-W. see Jaffe, M.: Vol. 117, pp. 297-328.
Chow, T. S.: Glassy State Relaxation and Deformation in Polymers. Vol. 103, pp. 149-190.
Chung, T.-S. see Jaffe, M.: Vol. 117, pp. 297-328.
Cölfen, H. and *Antonietti, M.*: Field-Flow Fractionation Techniques for Polymer and Colloid Analysis. Vol. 150, pp. 67-187.
Comanita, B. see Roovers, J.: Vol. 142, pp. 179-228.
Connell, J. W. see Hergenrother, P. M.: Vol. 117, pp. 67-110.
Creton, C., Kramer, E. J., Brown, H. R., Hui, C.-Y.: Adhesion and Fracture of Interfaces Between Immiscible Polymers: From the Molecular to the Continuum Scale. Vol. 156, pp. 53-135.
Criado-Sancho, M. see Jou, D.: Vol. 120, pp. 207-266.
Curro, J.G. see Schweizer, K.S.: Vol. 116, pp. 319-378.
Curtiss, C. F. and *Bird, R. B.*: Statistical Mechanics of Transport Phenomena: Polymeric Liquid Mixtures. Vol. 125, pp. 1-102.
Cussler, E. L., Wang, K. L. and *Burban, J. H.*: Hydrogels as Separation Agents. Vol. 110, pp. 67-80.

Dalton, L. Nonlinear Optical Polymeric Materials: From Chromophore Design to Commercial Applications. Vol. 158, pp. 1-86.
Davidson, J.M. see Prokop, A.: Vol. 160, pp.119-174.
Davis, K.A. see Matyjaszewski, K.: Vol. 159, pp.1-169.
DeSimone, J. M. see Canelas D. A.: Vol. 133, pp. 103-140.
DiMari, S. see Prokop, A.: Vol. 136, pp. 1-52.
Dimonie, M. V. see Hunkeler, D.: Vol. 112, pp. 115-134.
Dingenouts, N., Bolze, J., Pötschke, D., Ballauf, M.: Analysis of Polymer Latexes by Small-Angle X-Ray Scattering. Vol. 144, pp. 1-48.
Dodd, L. R. and *Theodorou, D. N.*: Atomistic Monte Carlo Simulation and Continuum Mean Field Theory of the Structure and Equation of State Properties of Alkane and Polymer Melts. Vol. 116, pp. 249-282.
Doelker, E.: Cellulose Derivatives. Vol. 107, pp. 199-266.
Dolden, J. G.: Calculation of a Mesogenic Index with Emphasis Upon LC-Polyimides. Vol. 141, pp. 189-245.
Domb, A. J., Amselem, S., Shah, J. and *Maniar, M.*: Polyanhydrides: Synthesis and Characterization. Vol.107, pp. 93-142.
Domb, A.J. see Kumar, M.N.V.R.: Vol. 160, pp. 45-118.
Doruker, P. see Baschnagel, J.: Vol. 152, pp. 41-156.
Dubois, P. see Mecerreyes, D.: Vol. 147, pp. 1-60.
Dubrovskii, S. A. see Kazanskii, K. S.: Vol. 104, pp. 97-134.
Dunkin, I. R. see Steinke, J.: Vol. 123, pp. 81-126.
Dunson, D. L. see McGrath, J. E.: Vol. 140, pp. 61-106.

Eastmond, G. C.: Poly(ε-caprolactone) Blends. Vol.149, pp. 59-223.
Economy, J. and *Goranov, K.*: Thermotropic Liquid Crystalline Polymers for High Performance Applications. Vol. 117, pp. 221-256.

Ediger, M. D. and *Adolf, D. B.*: Brownian Dynamics Simulations of Local Polymer Dynamics. Vol. 116, pp. 73-110.
Edlund, U. Albertsson, A.-C.: Degradable Polymer Microspheres for Controlled Drug Delivery. Vol. 157, pp. 53-98.
Edwards, S. F. see Aharoni, S. M.: Vol. 118, pp. 1-231.
Endo, T. see Yagci, Y.: Vol. 127, pp. 59-86.
Engelhardt, H. and *Grosche, O.*: Capillary Electrophoresis in Polymer Analysis. Vol. 150, pp. 189-217.
Erman, B. see Bahar, I.: Vol. 116, pp. 145-206.
Ewen, B, Richter, D.: Neutron Spin Echo Investigations on the Segmental Dynamics of Polymers in Melts, Networks and Solutions. Vol. 134, pp. 1-130.
Ezquerra, T. A. see Baltá-Calleja, F. J.: Vol. 108, pp. 1-48.

Faust, R. see Charleux, B: Vol. 142, pp. 1-70.
Fekete, E see Pukánszky, B: Vol. 139, pp. 109-154.
Fendler, J.H.: Membrane-Mimetic Approach to Advanced Materials. Vol. 113, pp. 1-209.
Fetters, L. J. see Xu, Z.: Vol. 120, pp. 1-50.
Förster, S. and *Schmidt, M.*: Polyelectrolytes in Solution. Vol. 120, pp. 51-134.
Freire, J. J.: Conformational Properties of Branched Polymers: Theory and Simulations. Vol. 143, pp. 35-112.
Frenkel, S. Y. see Bronnikov, S. V.: Vol. 125, pp. 103-146.
Frick, B. see Baltá-Calleja, F. J.: Vol. 108, pp. 1-48.
Fridman, M. L.: see Terent´eva, J. P.: Vol. 101, pp. 29-64.
Fukui, K. see Otaigbe, J. U.: Vol. 154, pp. 1-86.
Funke, W.: Microgels-Intramolecularly Crosslinked Macromolecules with a Globular Structure. Vol. 136, pp. 137-232.

Galina, H.: Mean-Field Kinetic Modeling of Polymerization: The Smoluchowski Coagulation Equation. Vol. 137, pp. 135-172.
Ganesh, K. see Kishore, K.: Vol. 121, pp. 81-122.
Gaw, K. O. and *Kakimoto, M.*: Polyimide-Epoxy Composites. Vol. 140, pp. 107-136.
Geckeler, K. E. see Rivas, B.: Vol. 102, pp. 171-188.
Geckeler, K. E.: Soluble Polymer Supports for Liquid-Phase Synthesis. Vol. 121, pp. 31-80.
Gehrke, S. H.: Synthesis, Equilibrium Swelling, Kinetics Permeability and Applications of Environmentally Responsive Gels. Vol. 110, pp. 81-144.
de Gennes, P.-G.: Flexible Polymers in Nanopores. Vol. 138, pp. 91-106.
Giannelis, E.P., Krishnamoorti, R., Manias, E.: Polymer-Silicate Nanocomposites: Model Systems for Confined Polymers and Polymer Brushes. Vol. 138, pp. 107-148.
Godovsky, D. Y.: Device Applications of Polymer-Nanocomposites. Vol. 153, pp. 163-205.
Godovsky, D. Y.: Electron Behavior and Magnetic Properties Polymer-Nanocomposites. Vol. 119, pp. 79-122.
González Arche, A. see Baltá-Calleja, F. J.: Vol. 108, pp. 1-48.
Goranov, K. see Economy, J.: Vol. 117, pp. 221-256.
Gramain, P. see Améduri, B.: Vol. 127, pp. 87-142.
Grest, G.S.: Normal and Shear Forces Between Polymer Brushes. Vol. 138, pp. 149-184.
Grigorescu, G, Kulicke, W.-M.: Prediction of Viscoelastic Properties and Shear Stability of Polymers in Solution. Vol. 152, p. 1-40.
Grosberg, A. and *Nechaev, S.*: Polymer Topology. Vol. 106, pp. 1-30.
Grosche, O. see Engelhardt, H.: Vol. 150, pp. 189-217.
Grubbs, R., Risse, W. and *Novac, B.*: The Development of Well-defined Catalysts for Ring-Opening Olefin Metathesis. Vol. 102, pp. 47-72.
Gubler, U., Bosshard, C.: Molecular Design for Third-Order Nonlinear Optics. Vol. 158, pp. 123-190.
van Gunsteren, W. F. see Gusev, A. A.: Vol. 116, pp. 207-248.
Gusev, A. A., Müller-Plathe, F., van Gunsteren, W. F. and *Suter, U. W.*: Dynamics of Small Molecules in Bulk Polymers. Vol. 116, pp. 207-248.

Gusev, A. A. see Baschnagel, J.: Vol. 152, pp. 41-156.
Guillot, J. see Hunkeler, D.: Vol. 112, pp. 115-134.
Guyot, A. and *Tauer, K.*: Reactive Surfactants in Emulsion Polymerization. Vol. 111, pp. 43-66.

Hadjichristidis, N., Pispas, S., Pitsikalis, M., Iatrou, H., Vlahos, C.: Asymmetric Star Polymers Synthesis and Properties. Vol. 142, pp. 71-128.
Hadjichristidis, N. see Xu, Z.: Vol. 120, pp. 1-50.
Hadjichristidis, N. see Pitsikalis, M.: Vol. 135, pp. 1-138.
Hahn, O. see Baschnagel, J.: Vol. 152, pp. 41-156.
Hakkarainen, M.: Aliphatic Polyesters: Abiotic and Biotic Degradation and Degradation Products. Vol. 157, pp. 1-26.
Hall, H. K. see *Penelle, J.*: Vol. 102, pp. 73-104.
Hamley, I. W.: Crystallization in Block Copolymers. Vol. 148, pp. 113-138.
Hammouda, B.: SANS from Homogeneous Polymer Mixtures: A Unified Overview. Vol. 106, pp. 87-134.
Han, M.J. and Chang, J.Y.: Polynucleotide Analogues. Vol. 153, pp. 1-36.
Harada, A.: Design and Construction of Supramolecular Architectures Consisting of Cyclodextrins and Polymers. Vol. 133, pp. 141-192.
Haralson, M. A. see Prokop, A.: Vol. 136, pp. 1-52.
Hassan, C.M. and Peppas, N.A.: Structure and Applications of Poly(vinyl alcohol) Hydrogels Produced by Conventional Crosslinking or by Freezing/Thawing Methods. Vol. 153, pp. 37-65.
Hawker, C. J. Dentritic and Hyperbranched Macromolecules – Precisely Controlled Macromolecular Architectures. Vol. 147, pp. 113-160.
Hawker, C. J. see Hedrick, J. L.: Vol. 141, pp. 1-44.
Hedrick, J. L., Carter, K. R., Labadie, J. W., Miller, R. D., Volksen, W., Hawker, C. J., Yoon, D. Y., Russell, T. P., McGrath, J. E., Briber, R. M.: Nanoporous Polyimides. Vol. 141, pp. 1-44.
Hedrick, J. L., Labadie, J. W., Volksen, W. and *Hilborn, J. G.*: Nanoscopically Engineered Polyimides. Vol. 147, pp. 61-112.
Hedrick, J. L. see Hergenrother, P. M.: Vol. 117, pp. 67-110.
Hedrick, J. L. see Kiefer, J.: Vol. 147, pp. 161-247.
Hedrick, J.L. see McGrath, J. E.: Vol. 140, pp. 61-106.
Heinrich, G. and Klüppel, M.: Recent Advances in the Theory of Filler Networking in Elastomers. Vol. 160, pp. 1-44.
Heller, J.: Poly (Ortho Esters). Vol. 107, pp. 41-92.
Hemielec, A. A. see Hunkeler, D.: Vol. 112, pp. 115-134.
Hergenrother, P. M., Connell, J. W., Labadie, J. W. and *Hedrick, J. L.*: Poly(arylene ether)s Containing Heterocyclic Units. Vol. 117, pp. 67-110.
Hernández-Barajas, J. see Wandrey, C.: Vol. 145, pp. 123-182.
Hervet, H. see Léger, L.: Vol. 138, pp. 185-226.
Hilborn, J. G. see Hedrick, J. L.: Vol. 147, pp. 61-112.
Hilborn, J. G. see Kiefer, J.: Vol. 147, pp. 161-247.
Hiramatsu, N. see Matsushige, M.: Vol. 125, pp. 147-186.
Hirasa, O. see Suzuki, M.: Vol. 110, pp. 241-262.
Hirotsu, S.: Coexistence of Phases and the Nature of First-Order Transition in Poly-N-isopropylacrylamide Gels. Vol. 110, pp. 1-26.
Höcker, H. see Klee, D.: Vol. 149, pp. 1-57.
Hornsby, P.: Rheology, Compoundind and Processing of Filled Thermoplastics. Vol. 139, pp. 155-216.
Hui, C.-Y. see Creton, C.: Vol. 156, pp. 53-135
Hult, A., Johansson, M., Malmström, E.: Hyperbranched Polymers. Vol. 143, pp. 1-34.
Hunkeler, D., Candau, F., Pichot, C., Hemielec, A. E., Xie, T. Y., Barton, J., Vaskova, V., Guillot, J., Dimonie, M. V., Reichert, K. H.: Heterophase Polymerization: A Physical and Kinetic Comparision and Categorization. Vol. 112, pp. 115-134.

Hunkeler, D. see Prokop, A.: Vol. 136, pp. 1-52; 53-74.
Hunkeler, D see Wandrey, C.: Vol. 145, pp. 123-182.

Iatrou, H. see Hadjichristidis, N.: Vol. 142, pp. 71-128.
Ichikawa, T. see Yoshida, H.: Vol. 105, pp. 3-36.
Ihara, E. see Yasuda, H.: Vol. 133, pp. 53-102.
Ikada, Y. see Uyama, Y.: Vol. 137, pp. 1-40.
Ilavsky, M.: Effect on Phase Transition on Swelling and Mechanical Behavior of Synthetic Hydrogels. Vol. 109, pp. 173-206.
Imai, Y.: Rapid Synthesis of Polyimides from Nylon-Salt Monomers. Vol. 140, pp. 1-23.
Inomata, H. see Saito, S.: Vol. 106, pp. 207-232.
Inoue, S. see Sugimoto, H.: Vol. 146, pp. 39-120.
Irie, M.: Stimuli-Responsive Poly(N-isopropylacrylamide), Photo- and Chemical-Induced Phase Transitions. Vol. 110, pp. 49-66.
Ise, N. see Matsuoka, H.: Vol. 114, pp. 187-232.
Ito, K., Kawaguchi, S,:Poly(macronomers), Homo- and Copolymerization. Vol. 142, pp. 129-178.
Ivanov, A. E. see Zubov, V. P.: Vol. 104, pp. 135-176.

Jacob, S. and Kennedy, J.: Synthesis, Characterization and Properties of OCTA-ARM Polyisobutylene-Based Star Polymers. Vol. 146, pp. 1-38.
Jaffe, M., Chen, P., Choe, E.-W., Chung, T.-S. and *Makhija, S.*: High Performance Polymer Blends. Vol. 117, pp. 297-328.
Jancar, J.: Structure-Property Relationships in Thermoplastic Matrices. Vol. 139, pp. 1-66.
Jerôme, R.: see Mecerreyes, D.: Vol. 147, pp. 1-60.
Jiang, M., Li, M., Xiang, M. and Zhou, H.: Interpolymer Complexation and Miscibility and Enhancement by Hydrogen Bonding. Vol. 146, pp. 121-194.
Jin, J.: see Shim, H.-K.: Vol. 158, pp. 191-241.
Jo, W. H. and Yang, J. S.: Molecular Simulation Approaches for Multiphase Polymer Systems. Vol. 156, pp. 1-52.
Johansson, M. see Hult, A.: Vol. 143, pp. 1-34.
Joos-Müller, B. see Funke, W.: Vol. 136, pp. 137-232.
Jou, D., Casas-Vazquez, J. and Criado-Sancho, M.: Thermodynamics of Polymer Solutions under Flow: Phase Separation and Polymer Degradation. Vol. 120, pp. 207-266.

Kaetsu, I.: Radiation Synthesis of Polymeric Materials for Biomedical and Biochemical Applications. Vol. 105, pp. 81-98.
Kaji, K. see Kanaya, T.: Vol. 154, pp. 87-141.
Kakimoto, M. see Gaw, K. O.: Vol. 140, pp. 107-136.
Kaminski, W. and *Arndt, M.*: Metallocenes for Polymer Catalysis. Vol. 127, pp. 143-187.
Kammer, H. W., Kressler, H. and *Kummerloewe, C.*: Phase Behavior of Polymer Blends - Effects of Thermodynamics and Rheology. Vol. 106, pp. 31-86.
Kanaya, T. and Kaji, K.: Dynamcis in the Glassy State and Near the Glass Transition of Amorphous Polymers as Studied by Neutron Scattering. Vol. 154, pp. 87-141.
Kandyrin, L. B. and *Kuleznev, V. N.*: The Dependence of Viscosity on the Composition of Concentrated Dispersions and the Free Volume Concept of Disperse Systems. Vol. 103, pp. 103-148.
Kaneko, M. see Ramaraj, R.: Vol. 123, pp. 215-242.
Kang, E. T., Neoh, K. G. and *Tan, K. L.*: X-Ray Photoelectron Spectroscopic Studies of Electroactive Polymers. Vol. 106, pp. 135-190.
Karlsson, S. see Söderqvist Lindblad, M.: Vol. 157, pp. 139–161.
Kato, K. see Uyama, Y.: Vol. 137, pp. 1-40.
Kawaguchi, S. see Ito, K.: Vol. 142, p 129-178.
Kazanskii, K. S. and *Dubrovskii, S. A.*: Chemistry and Physics of „Agricultural" Hydrogels. Vol. 104, pp. 97-134.
Kennedy, J. P. see Jacob, S.: Vol. 146, pp. 1-38.
Kennedy, J. P. see Majoros, I.: Vol. 112, pp. 1-113.

Khokhlov, A., Starodybtzev, S. and *Vasilevskaya, V*.: Conformational Transitions of Polymer Gels: Theory and Experiment. Vol. 109, pp. 121-172.
Kiefer, J., Hedrick J. L. and *Hiborn, J. G.*: Macroporous Thermosets by Chemically Induced Phase Separation. Vol. 147, pp. 161-247.
Kilian, H. G. and *Pieper, T.*: Packing of Chain Segments. A Method for Describing X-Ray Patterns of Crystalline, Liquid Crystalline and Non-Crystalline Polymers. Vol. 108, pp. 49-90.
Kim, J. see Quirk, R.P.: Vol. 153, pp. 67-162.
Kishore, K. and *Ganesh, K.*: Polymers Containing Disulfide, Tetrasulfide, Diselenide and Ditelluride Linkages in the Main Chain. Vol. 121, pp. 81-122.
Kitamaru, R.: Phase Structure of Polyethylene and Other Crystalline Polymers by Solid-State ^{13}C/MNR. Vol. 137, pp 41-102.
Klee, D. and *Höcker, H.*: Polymers for Biomedical Applications: Improvement of the Interface Compatibility. Vol. 149, pp. 1-57.
Klier, J. see Scranton, A. B.: Vol. 122, pp. 1-54.
Klüppel, M. see Heinrich, G.: Vol. 160, pp 1-44.
Kobayashi, S., Shoda, S. and *Uyama, H.*: Enzymatic Polymerization and Oligomerization. Vol. 121, pp. 1-30.
Köhler, W. and *Schäfer, R.*: Polymer Analysis by Thermal-Diffusion Forced Rayleigh Scattering. Vol. 151, pp. 1-59.
Koenig, J. L. see Andreis, M.: Vol. 124, pp. 191-238.
Koike, T.: Viscoelastic Behavior of Epoxy Resins Before Crosslinking. Vol. 148, pp. 139-188.
Kokufuta, E.: Novel Applications for Stimulus-Sensitive Polymer Gels in the Preparation of Functional Immobilized Biocatalysts. Vol. 110, pp. 157-178.
Konno, M. see Saito, S.: Vol. 109, pp. 207-232.
Kopecek, J. see Putnam, D.: Vol. 122, pp. 55-124.
Koßmehl, G. see Schopf, G.: Vol. 129, pp. 1-145.
Kozlov, E. see Prokop, A.: Vol. 160, pp. 119-174.
Kramer, E. J. see Creton, C.: Vol. 156, pp. 53-135.
Kremer, K. see Baschnagel, J.: Vol. 152, pp. 41-156.
Kressler, J. see Kammer, H. W.: Vol. 106, pp. 31-86.
Kricheldorf, H. R.: Liquid-Cristalline Polyimides. Vol. 141, pp. 83-188.
Krishnamoorti, R. see Giannelis, E.P.: Vol. 138, pp. 107-148.
Kirchhoff, R. A. and *Bruza, K. J.*: Polymers from Benzocyclobutenes. Vol. 117, pp. 1-66.
Kuchanov, S. I.: Modern Aspects of Quantitative Theory of Free-Radical Copolymerization. Vol. 103, pp. 1-102.
Kuchanov, S. I.: Principles of Quantitive Description of Chemical Structure of Synthetic Polymers. Vol. 152, p. 157-202.
Kudaibergennow, S.E.: Recent Advances in Studying of Synthetic Polyampholytes in Solutions. Vol. 144, pp. 115-198.
Kuleznev, V. N. see Kandyrin, L. B.: Vol. 103, pp. 103-148.
Kulichkhin, S. G. see Malkin, A. Y.: Vol. 101, pp. 217-258.
Kulicke, W.-M. see Grigorescu, G.: Vol. 152, p. 1-40.
Kumar, M.N.V.R., Kumar, N., Domb, A.J. and *Arora, M.*: Pharmaceutical Polyme-ric Controlled Drug Delivery Systems. Vol. 160, pp. 45-118.
Kumar, N. see Kumar M.N.V.R.: Vol. 160, pp. 45-118.
Kummerloewe, C. see Kammer, H. W.: Vol. 106, pp. 31-86.
Kuznetsova, N. P. see Samsonov, G. V.: Vol. 104, pp. 1-50.Labadie, J. W. see Hergenrother, P. M.: Vol. 117, pp. 67-110.

Labadie, J. W. see Hedrick, J. L.: Vol. 141, pp. 1-44.
Labadie, J. W. see Hedrick, J. L.: Vol. 147, pp. 61-112.
Lamparski, H. G. see O´Brien, D. F.: Vol. 126, pp. 53-84.
Laschewsky, A.: Molecular Concepts, Self-Organisation and Properties of Polysoaps. Vol. 124, pp. 1-86.
Laso, M. see Leontidis, E.: Vol. 116, pp. 283-318.

Lazár, M. and *Rychlŏ, R.*: Oxidation of Hydrocarbon Polymers. Vol. 102, pp. 189-222.
Lechowicz, J. see Galina, H.: Vol. 137, pp. 135-172.
Léger, L., Raphaël, E., Hervet, H.: Surface-Anchored Polymer Chains: Their Role in Adhesion and Friction. Vol. 138, pp. 185-226.
Lenz, R. W.: Biodegradable Polymers. Vol. 107, pp. 1-40.
Leontidis, E., de Pablo, J. J., Laso, M. and *Suter, U. W.*: A Critical Evaluation of Novel Algorithms for the Off-Lattice Monte Carlo Simulation of Condensed Polymer Phases. Vol. 116, pp. 283-318.
Lee, B. see Quirk, R.P: Vol. 153, pp. 67-162.
Lee, Y. see Quirk, R.P: Vol. 153, pp. 67-162.
Lesec, J. see Viovy, J.-L.: Vol. 114, pp. 1-42.
Li, M. see Jiang, M.: Vol. 146, pp. 121-194.
Liang, G. L. see Sumpter, B. G.: Vol. 116, pp. 27-72.
Lienert, K.-W.: Poly(ester-imide)s for Industrial Use. Vol. 141, pp. 45-82.
Lin, J. and *Sherrington, D. C.*: Recent Developments in the Synthesis, Thermostability and Liquid Crystal Properties of Aromatic Polyamides. Vol. 111, pp. 177-220.
Liu, Y. see Söderqvist Lindblad, M.: Vol. 157, pp. 139–161
López Cabarcos, E. see Baltá-Calleja, F. J.: Vol. 108, pp. 1-48.

Majoros, I., Nagy, A. and *Kennedy, J. P.*: Conventional and Living Carbocationic Polymerizations United. I. A Comprehensive Model and New Diagnostic Method to Probe the Mechanism of Homopolymerizations. Vol. 112, pp. 1-113.
Makhija, S. see Jaffe, M.: Vol. 117, pp. 297-328.
Malmström, E. see Hult, A.: Vol. 143, pp. 1-34.
Malkin, A. Y. and *Kulichkhin, S. G.*: Rheokinetics of Curing. Vol. 101, pp. 217-258.
Maniar, M. see Domb, A. J.: Vol. 107, pp. 93-142.
Manias, E., see Giannelis, E.P.: Vol. 138, pp. 107-148.
Mashima, K., Nakayama, Y. and *Nakamura, A.*: Recent Trends in Polymerization of a-Olefins Catalyzed by Organometallic Complexes of Early Transition Metals. Vol. 133, pp. 1-52.
Mathew, D. see Reghunadhan Nair, C.P.: Vol. 155, pp. 1-99.
Matsumoto, A.: Free-Radical Crosslinking Polymerization and Copolymerization of Multivinyl Compounds. Vol. 123, pp. 41-80.
Matsumoto, A. see Otsu, T.: Vol. 136, pp. 75-138.
Matsuoka, H. and *Ise, N.*: Small-Angle and Ultra-Small Angle Scattering Study of the Ordered Structure in Polyelectrolyte Solutions and Colloidal Dispersions. Vol. 114, pp. 187-232.
Matsushige, K., Hiramatsu, N. and *Okabe, H.*: Ultrasonic Spectroscopy for Polymeric Materials. Vol. 125, pp. 147-186.
Mattice, W. L. see Rehahn, M.: Vol. 131/132, pp. 1-475.
Mattice, W. L. see Baschnagel, J.: Vol. 152, p. 41-156.
Matyjaszewski, K., Davis, K.A.: Statistical, Gradient, Block and Graft Copolymers by Controlled/Living Radical Polymerizations. Vol. 159, p. 1-169.
Mays, W. see Xu, Z.: Vol. 120, pp. 1-50.
Mays, J.W. see Pitsikalis, M.: Vol.135, pp. 1-138.
McGrath, J. E. see Hedrick, J. L.: Vol. 141, pp. 1-44.
McGrath, J. E., Dunson, D. L., Hedrick, J. L.: Synthesis and Characterization of Segmented Polyimide-Polyorganosiloxane Copolymers. Vol. 140, pp. 61-106.
McLeish, T.C.B., Milner, S. T.: Entangled Dynamics and Melt Flow of Branched Polymers. Vol. 143, pp. 195-256.
Mecerreyes, D., Dubois, P. and *Jerôme, R.*: Novel Macromolecular Architectures Based on Aliphatic Polyesters: Relevance of the „Coordination-Insertion" Ring-Opening Polymerization. Vol. 147, pp. 1 -60.
Mecham, S. J. see McGrath, J. E.: Vol. 140, pp. 61-106.
Mikos, A. G. see Thomson, R. C.: Vol. 122, pp. 245-274.
Milner, S. T. see McLeish, T. C. B.: Vol. 143, pp. 195-256.
Mison, P. and Sillion, B.: Thermosetting Oligomers Containing Maleimides and Nadiimides End-Groups. Vol. 140, pp. 137-180.
Miyasaka, K.: PVA-Iodine Complexes: Formation, Structure and Properties. Vol. 108. pp. 91-130.

Miller, R. D. see Hedrick, J. L.: Vol. 141, pp. 1-44.
Monnerie, L. see Bahar, I.: Vol. 116, pp. 145-206.
Morishima, Y.: Photoinduced Electron Transfer in Amphiphilic Polyelectrolyte Systems. Vol. 104, pp. 51-96.
Morton M. see Quirk, R.P: Vol. 153, pp. 67-162
Mours, M. see Winter, H. H.: Vol. 134, pp. 165-234.
Müllen, K. see Scherf, U.: Vol. 123, pp. 1-40.
Müller-Plathe, F. see Gusev, A. A.: Vol. 116, pp. 207-248.
Müller-Plathe, F. see Baschnagel, J.: Vol. 152, p. 41-156.
Mukerherjee, A. see Biswas, M.: Vol. 115, pp. 89-124.
Murat, M. see Baschnagel, J.: Vol. 152, p. 41-156.
Mylnikov, V.: Photoconducting Polymers. Vol. 115, pp. 1-88.

Nagy, A. see Majoros, I.: Vol. 112, pp. 1-11.
Nakamura, A. see Mashima, K.: Vol. 133, pp. 1-52.
Nakayama, Y. see Mashima, K.: Vol. 133, pp. 1-52.
Narasinham, B., Peppas, N. A.: The Physics of Polymer Dissolution: Modeling Approaches and Experimental Behavior. Vol. 128, pp. 157-208.
Nechaev, S. see Grosberg, A.: Vol. 106, pp. 1-30.
Neoh, K. G. see Kang, E. T.: Vol. 106, pp. 135-190.
Newman, S. M. see Anseth, K. S.: Vol. 122, pp. 177-218.
Nijenhuis, K. te: Thermoreversible Networks. Vol. 130, pp. 1-252.
Ninan, K.N. see Reghunadhan Nair, C. P.: Vol. 155, pp. 1-99.
Noid, D. W. see Otaigbe, J.U.: Vol. 154, pp. 1-86.
Noid, D. W. see Sumpter, B. G.: Vol. 116, pp. 27-72.
Novac, B. see Grubbs, R.: Vol. 102, pp. 47-72.
Novikov, V. V. see Privalko, V. P.: Vol. 119, pp. 31-78.

O'Brien, D. F., Armitage, B. A., Bennett, D. E. and *Lamparski, H. G.*: Polymerization and Domain Formation in Lipid Assemblies. Vol. 126, pp. 53-84.
Ogasawara, M.: Application of Pulse Radiolysis to the Study of Polymers and Polymerizations. Vol.105, pp. 37-80.
Okabe, H. see Matsushige, K.: Vol. 125, pp. 147-186.
Okada, M.: Ring-Opening Polymerization of Bicyclic and Spiro Compounds. Reactivities and Polymerization Mechanisms. Vol. 102, pp. 1-46.
Okano, T.: Molecular Design of Temperature-Responsive Polymers as Intelligent Materials. Vol. 110, pp. 179-198.
Okay, O. see Funke, W.: Vol. 136, pp. 137-232.
Onuki, A.: Theory of Phase Transition in Polymer Gels. Vol. 109, pp. 63-120.
Osad'ko, I.S.: Selective Spectroscopy of Chromophore Doped Polymers and Glasses. Vol. 114, pp. 123-186.
Otaigbe, J. U., Barnes, M. D., Fukui, K., Sumpter, B. G., Noid, D. W.: Generation, Characterization, and Modeling of Polymer Micro- and Nano-Particles. Vol. 154, pp. 1-86.
Otsu, T., Matsumoto, A.: Controlled Synthesis of Polymers Using the Iniferter Technique: Developments in Living Radical Polymerization. Vol. 136, pp. 75-138.

de Pablo, J. J. see Leontidis, E.: Vol. 116, pp. 283-318.
Padias, A. B. see Penelle, J.: Vol. 102, pp. 73-104.
Pascault, J.-P. see Williams, R. J. J.: Vol. 128, pp. 95-156.
Pasch, H.: Analysis of Complex Polymers by Interaction Chromatography. Vol. 128, pp. 1-46.
Pasch, H.: Hyphenated Techniques in Liquid Chromatography of Polymers. Vol. 150, pp. 1-66.
Paul, W. see Baschnagel, J.: Vol. 152, p. 41-156.
Penczek, P. see Batog, A. E.: Vol. 144, pp. 49-114.
Penelle, J., Hall, H. K., Padias, A. B. and *Tanaka, H.*: Captodative Olefins in Polymer Chemistry. Vol. 102, pp. 73-104.

Peppas, N. A. see Bell, C. L.: Vol. 122, pp. 125-176.
Peppas, N.A. see Hassan, C.M.: Vol. 153, pp. 37-65
Peppas, N. A. see Narasimhan, B.: Vol. 128, pp. 157-208.
Pet'ko, I. P. see *Batog, A. E.:* Vol. 144, pp. 49-114.
Pichot, C. see Hunkeler, D.: Vol. 112, pp. 115-134.
Pieper, T. see Kilian, H. G.: Vol. 108, pp. 49-90.
Pispas, S. see Pitsikalis, M.: Vol. 135, pp. 1-138.
Pispas, S. see Hadjichristidis: Vol. 142, pp. 71-128.
Pitsikalis, M., Pispas, S., Mays, J. W., Hadjichristidis, N.: Nonlinear Block Copolymer Architectures. Vol. 135, pp. 1-138.
Pitsikalis, M. see Hadjichristidis: Vol. 142, pp. 71-128.
Pötschke, D. see Dingenouts, N.: Vol 144, pp. 1-48.
Pokrovskii, V. N.: The Mesoscopic Theory of the Slow Relaxation of Linear Macromolecules. Vol. 154, pp. 143-219.
Pospíšil, J.: Functionalized Oligomers and Polymers as Stabilizers for Conventional Polymers. Vol. 101, pp. 65-168.
Pospíšil, J.: Aromatic and Heterocyclic Amines in Polymer Stabilization. Vol. 124, pp. 87-190.
Powers, A. C. see Prokop, A.: Vol. 136, pp. 53-74.
Priddy, D. B.: Recent Advances in Styrene Polymerization. Vol. 111, pp. 67-114.
Priddy, D. B.: Thermal Discoloration Chemistry of Styrene-co-Acrylonitrile. Vol. 121, pp. 123-154.
Privalko, V. P. and *Novikov, V. V.:* Model Treatments of the Heat Conductivity of Heterogeneous Polymers. Vol. 119, pp 31-78.
Prokop, A., Hunkeler, D., Powers, A. C., Whitesell, R. R., Wang, T. G.: Water Soluble Polymers for Immunoisolation II: Evaluation of Multicomponent Microencapsulation Systems. Vol. 136, pp. 53-74.
Prokop, A., Hunkeler, D., DiMari, S., Haralson, M. A., Wang, T. G.: Water Soluble Polymers for Immunoisolation I: Complex Coacervation and Cytotoxicity. Vol. 136, pp. 1-52.
Prokop, A., Kozlov, E., Carlesso, G. and Davidsen, J.M.: Hydrogel-Based Colloidal Polymeric System for Protein and Drug Delivery: Physical and Chemical Characterization, Permeability Control and Applications. Vol. 160, pp. 119-174.
Pukánszky, B. and *Fekete, E.:* Adhesion and Surface Modification. Vol. 139, pp. 109-154.
Putnam, D. and *Kopecek, J.:* Polymer Conjugates with Anticancer Acitivity. Vol. 122, pp. 55-124.

Quirk, R.P. and Yoo, T., Lee, Y., M., Kim, J. and Lee, B.: Applications of 1,1-Diphenylethylene Chemistry in Anionic Synthesis of Polymers with Controlled Structures. Vol. 153, pp. 67-162.

Ramaraj, R. and *Kaneko, M.:* Metal Complex in Polymer Membrane as a Model for Photosynthetic Oxygen Evolving Center. Vol. 123, pp. 215-242.
Rangarajan, B. see Scranton, A. B.: Vol. 122, pp. 1-54.
Ranucci, E. see Söderqvist Lindblad, M.: Vol. 157, pp. 139–161.
Raphaël, E. see Léger, L.: Vol. 138, pp. 185-226.
Reddinger, J. L. and *Reynolds, J. R.:* Molecular Engineering of π-Conjugated Polymers. Vol. 145, pp. 57-122.
Reghunadhan Nair, C.P., Mathew, D. and *Ninan, K.N.,* : Cyanate Ester Resins, Recent Developments. Vol. 155, pp. 1-99.
Reichert, K. H. see Hunkeler, D.: Vol. 112, pp. 115-134.
Rehahn, M., Mattice, W. L., Suter, U. W.: Rotational Isomeric State Models in Macromolecular Systems. Vol. 131/132, pp. 1-475.
Reynolds, J.R. see Reddinger, J. L.: Vol. 145, pp. 57-122.
Richter, D. see Ewen, B.: Vol. 134, pp.1-130.
Risse, W. see Grubbs, R.: Vol. 102, pp. 47-72.
Rivas, B. L. and *Geckeler, K. E.:* Synthesis and Metal Complexation of Poly(ethyleneimine) and Derivatives. Vol. 102, pp. 171-188.
Robin, J. J. see Boutevin, B.: Vol. 102, pp. 105-132.

Roe, R.-J.: MD Simulation Study of Glass Transition and Short Time Dynamics in Polymer Liquids. Vol. 116, pp. 111-114.
Roovers, J., Comanita, B.: Dendrimers and Dendrimer-Polymer Hybrids. Vol. 142, pp 179-228.
Rothon, R. N.: Mineral Fillers in Thermoplastics: Filler Manufacture and Characterisation. Vol. 139, pp. 67-108.
Rozenberg, B. A. see Williams, R. J. J.: Vol. 128, pp. 95-156.
Ruckenstein, E.: Concentrated Emulsion Polymerization. Vol. 127, pp. 1-58.
Rusanov, A. L.: Novel Bis (Naphtalic Anhydrides) and Their Polyheteroarylenes with Improved Processability. Vol. 111, pp. 115-176.
Russel, T. P. see Hedrick, J. L.: Vol. 141, pp. 1-44.
Rychlý, J. see Lazár, M.: Vol. 102, pp. 189-222.
Ryner, M. see Stridsberg, K. M.: Vol. 157, pp. 27–51.
Ryzhov, V. A. see Bershtein, V. A.: Vol. 114, pp. 43-122.

Sabsai, O. Y. see Barshtein, G. R.: Vol. 101, pp. 1-28.
Saburov, V. V. see Zubov, V. P.: Vol. 104, pp. 135-176.
Saito, S., Konno, M. and *Inomata, H.*: Volume Phase Transition of N-Alkylacrylamide Gels. Vol. 109, pp. 207-232.
Samsonov, G. V. and *Kuznetsova, N. P.*: Crosslinked Polyelectrolytes in Biology. Vol. 104, pp. 1-50.
Santa Cruz, C. see Baltá-Calleja, F. J.: Vol. 108, pp. 1-48.
Santos, S. see Baschnagel, J.: Vol. 152, p. 41-156.
Sato, T. and *Teramoto, A.*: Concentrated Solutions of Liquid-Christalline Polymers. Vol. 126, pp. 85-162.
Schäfer R. see Köhler, W.: Vol. 151, pp. 1-59.
Scherf, U. and *Müllen, K.*: The Synthesis of Ladder Polymers. Vol. 123, pp. 1-40.
Schmidt, M. see Förster, S.: Vol. 120, pp. 51-134.
Schopf, G. and *Koßmehl, G.*: Polythiophenes - Electrically Conductive Polymers. Vol. 129, pp. 1-145.
Schweizer, K. S.: Prism Theory of the Structure, Thermodynamics, and Phase Transitions of Polymer Liquids and Alloys. Vol. 116, pp. 319-378.
Scranton, A. B., Rangarajan, B. and *Klier, J.*: Biomedical Applications of Polyelectrolytes. Vol. 122, pp. 1-54.
Sefton, M. V. and *Stevenson, W. T. K.*: Microencapsulation of Live Animal Cells Using Polycrylates. Vol. 107, pp. 143-198.
Shamanin, V. V.: Bases of the Axiomatic Theory of Addition Polymerization. Vol. 112, pp. 135-180.
Sheiko, S. S.: Imaging of Polymers Using Scanning Force Microscopy: From Superstructures to Individual Molecules. Vol. 151, pp. 61-174.
Sherrington, D. C. see Cameron, N. R. , Vol. 126, pp. 163-214.
Sherrington, D. C. see Lin, J.: Vol. 111, pp. 177-220.
Sherrington, D. C. see Steinke, J.: Vol. 123, pp. 81-126.
Shibayama, M. see Tanaka, T.: Vol. 109, pp. 1-62.
Shiga, T.: Deformation and Viscoelastic Behavior of Polymer Gels in Electric Fields. Vol. 134, pp. 131-164.
Shim, H.-K., Jin, J.: Light-Emitting Characteristics of Conjugated Polymers. Vol. 158, pp. 191-241.
Shoda, S. see Kobayashi, S.: Vol. 121, pp. 1-30.
Siegel, R. A.: Hydrophobic Weak Polyelectrolyte Gels: Studies of Swelling Equilibria and Kinetics. Vol. 109, pp. 233-268.
Silvestre, F. see Calmon-Decriaud, A.: Vol. 207, pp. 207-226.
Sillion, B. see Mison, P.: Vol. 140, pp. 137-180.
Singh, R. P. see Sivaram, S.: Vol. 101, pp. 169-216.
Sinha Ray, S. see Biswas, M: Vol. 155, pp. 167-221.
Sivaram, S. and *Singh, R. P.*: Degradation and Stabilization of Ethylene-Propylene Copolymers and Their Blends: A Critical Review. Vol. 101, pp. 169-216.

Söderqvist Lindblad, M., Liu, Y., Albertsson, A.-C., Ranucci, E., Karlsson, S.: Polymer from Renewable Resources. Vol. 157, pp. 139-161
Starodybtzev, S. see Khokhlov, A.: Vol. 109, pp. 121-172.
Stegeman, G. I.: see Canva, M.: Vol. 158, pp. 87-121.
Steinke, J., Sherrington, D. C. and *Dunkin, I. R.*: Imprinting of Synthetic Polymers Using Molecular Templates. Vol. 123, pp. 81-126.
Stenzenberger, H. D.: Addition Polyimides. Vol. 117, pp. 165-220.
Stevenson, W. T. K. see Sefton, M. V.: Vol. 107, pp. 143-198.
Stridsberg, K. M., Ryner, M., Albertsson, A.-C.: Controlled Ring-Opening Polymerization: Polymers with Designed Macromoleculars Architecture. Vol. 157, pp. 27-51.
Suematsu, K.: Recent Progress of Gel Theory: Ring, Excluded Volume, and Dimension. Vol. 156, pp. 136-214.
Sumpter, B. G., Noid, D. W., Liang, G. L. and *Wunderlich, B.*: Atomistic Dynamics of Macromolecular Crystals. Vol. 116, pp. 27-72.
Sumpter, B. G. see Otaigbe, J.U.: Vol. 154, pp. 1-86.
Sugimoto, H. and *Inoue, S.*: Polymerization by Metalloporphyrin and Related Complexes. Vol. 146, pp. 39-120.
Suter, U. W. see Gusev, A. A.: Vol. 116, pp. 207-248.
Suter, U. W. see Leontidis, E.: Vol. 116, pp. 283-318.
Suter, U. W. see Rehahn, M.: Vol. 131/132, pp. 1-475.
Suter, U. W. see Baschnagel, J.: Vol. 152, p. 41-156.
Suzuki, A.: Phase Transition in Gels of Sub-Millimeter Size Induced by Interaction with Stimuli. Vol. 110, pp. 199-240.
Suzuki, A. and *Hirasa, O.*: An Approach to Artifical Muscle by Polymer Gels due to Micro-Phase Separation. Vol. 110, pp. 241-262.

Tagawa, S.: Radiation Effects on Ion Beams on Polymers. Vol. 105, pp. 99-116.
Tan, K. L. see Kang, E. T.: Vol. 106, pp. 135-190.
Tanaka, H. and *Shibayama, M.*: Phase Transition and Related Phenomena of Polymer Gels. Vol. 109, pp. 1-62.
Tanaka, T. see Penelle, J.: Vol. 102, pp. 73-104.
Tauer, K. see Guyot, A.: Vol. 111, pp. 43-66.
Teramoto, A. see Sato, T.: Vol. 126, pp. 85-162.
Terent´eva, J. P. and *Fridman, M. L.*: Compositions Based on Aminoresins. Vol. 101, pp. 29-64.
Theodorou, D. N. see Dodd, L. R.: Vol. 116, pp. 249-282.
Thomson, R. C., Wake, M. C., Yaszemski, M. J. and *Mikos, A. G.*: Biodegradable Polymer Scaffolds to Regenerate Organs. Vol. 122, pp. 245-274.
Tokita, M.: Friction Between Polymer Networks of Gels and Solvent. Vol. 110, pp. 27-48.
Tries, V. see Baschnagel, J:. Vol. 152, p. 41-156.
Tsuruta, T.: Contemporary Topics in Polymeric Materials for Biomedical Applications. Vol. 126, pp. 1-52.

Uyama, H. see Kobayashi, S.: Vol. 121, pp. 1-30.
Uyama, Y: Surface Modification of Polymers by Grafting. Vol. 137, pp. 1-40.

Varma, I. K. see Albertsson, A.-C.: Vol. 157, pp. 99-138.
Vasilevskaya, V. see Khokhlov, A.: Vol. 109, pp. 121-172.
Vaskova, V. see Hunkeler, D.: Vol.:112, pp. 115-134.
Verdugo, P.: Polymer Gel Phase Transition in Condensation-Decondensation of Secretory Products. Vol. 110, pp. 145-156.
Vettegren, V. I.: see Bronnikov, S. V.: Vol. 125, pp. 103-146.
Viovy, J.-L. and *Lesec, J.*: Separation of Macromolecules in Gels: Permeation Chromatography and Electrophoresis. Vol. 114, pp. 1-42.
Vlahos, C. see Hadjichristidis, N.: Vol. 142, pp. 71-128.

Volksen, W.: Condensation Polyimides: Synthesis, Solution Behavior, and Imidization Characteristics. Vol. 117, pp. 111-164.
Volksen, W. see Hedrick, J. L.: Vol. 141, pp. 1-44.
Volksen, W. see Hedrick, J. L.: Vol. 147, pp. 61-112.

Wake, M. C. see Thomson, R. C.: Vol. 122, pp. 245-274.
Wandrey C., Hernández-Barajas, J. and *Hunkeler, D.*: Diallyldimethylammonium Chloride and its Polymers. Vol. 145, pp. 123-182.
Wang, K. L. see Cussler, E. L.: Vol. 110, pp. 67-80.
Wang, S.-Q.: Molecular Transitions and Dynamics at Polymer/Wall Interfaces: Origins of Flow Instabilities and Wall Slip. Vol. 138, pp. 227-276.
Wang, T. G. see Prokop, A.: Vol. 136, pp.1-52; 53-74.
Whitesell, R. R. see Prokop, A.: Vol. 136, pp. 53-74.
Williams, R. J. J., Rozenberg, B. A., Pascault, J.-P.: Reaction Induced Phase Separation in Modified Thermosetting Polymers. Vol. 128, pp. 95-156.
Winter, H. H., Mours, M.: Rheology of Polymers Near Liquid-Solid Transitions. Vol. 134, pp. 165-234.
Wu, C.: Laser Light Scattering Characterization of Special Intractable Macromolecules in Solution. Vol 137, pp. 103-134.
Wunderlich, B. see Sumpter, B. G.: Vol. 116, pp. 27-72.

Xiang, M. see Jiang, M.: Vol. 146, pp. 121-194.
Xie, T. Y. see Hunkeler, D.: Vol. 112, pp. 115-134.
Xu, Z., Hadjichristidis, N., Fetters, L. J. and *Mays, J. W.*: Structure/Chain-Flexibility Relationships of Polymers. Vol. 120, pp. 1-50.

Yagci, Y. and *Endo, T.*: N-Benzyl and N-Alkoxy Pyridium Salts as Thermal and Photochemical Initiators for Cationic Polymerization. Vol. 127, pp. 59-86.
Yannas, I. V.: Tissue Regeneration Templates Based on Collagen-Glycosaminoglycan Copolymers. Vol. 122, pp. 219-244.
Yang, J. S. see Jo, W. H.: Vol. 156, pp. 1-52.
Yamaoka, H.: Polymer Materials for Fusion Reactors. Vol. 105, pp. 117-144.
Yasuda, H. and *Ihara, E.*: Rare Earth Metal-Initiated Living Polymerizations of Polar and Nonpolar Monomers. Vol. 133, pp. 53-102.
Yaszemski, M. J. see Thomson, R. C.: Vol. 122, pp. 245-274.
Yoo, T. see Quirk, R.P.: Vol. 153, pp. 67-162.
Yoon, D. Y. see Hedrick, J. L.: Vol. 141, pp. 1-44.
Yoshida, H. and *Ichikawa, T.*: Electron Spin Studies of Free Radicals in Irradiated Polymers. Vol. 105, pp. 3-36.

Zhou, H. see Jiang, M.: Vol. 146, pp. 121-194.
Zubov, V. P., Ivanov, A. E. and *Saburov, V. V.*: Polymer-Coated Adsorbents for the Separation of Biopolymers and Particles. Vol. 104, pp. 135-176.

Subject Index

ABC triblock copolymers   56–58, 83–84, 101
AIBN   4, 37, 41, 69, 85, 108, 117, 119, 120, 123, 166,
A-T   31, 32, 36
ATRP   10–11, 12, 13
–, block copolymers via   44–68, 71, 73, 74, 75–76, 77, 78, 79–80, 81–82, 84–87, 89, 90–92, 93–95, 97–99, 100–103
–, equilibrium   45
–, graft copolymers via   109–113, 114–116, 118–119, 120–126
–, halogen exchange technique   45, 46, 47, 55, 56, 59, 64, 69, 71
–, reverse   53–54, 131
– sequential addition of monomers technique   46, 48, 49–50, 61, 71
–, star polymers via   138–142
–, statistical copolymers   19–27, 28, 65
p-Acetoxystyrene, copolymers with
–, styrene   15, 24
–, [(4'-methoxyphenyl)4-oxybenzoate]-6-hexyl (4-vinylbenzoate)   34
Acrylonitrile   38
Amphiphilic block copolymers   40, 42, 44, 56, 59–61, 63, 73, 90, 99–100, 130

Benzyl methacrylate   66
Biphasic ATRP   66, 71
Bottle-brush copolymer   114–116
BPPN   38, 43, 166
Butadiene
–, copolymers with methyl methacrylate   102–103
–, copolymers with styrene   39–40, 78, 96, 97, 102
–, functionalization with an ATRP initiator   78, 102

–, functionalization with a NMP initiator   96, 97
N-$tert$-Butyl acrylamide   61
n-Butyl acrylate, copolymers with
–, acrylic acid   37
–, t-butyl acrylate   37
–, N,N-dimethylacrylamide   37, 62
–, ethylene glycol monomethacrylate monoperfluorooctanoate   64
–, glycidyl acrylate   37
–, 2-hydroxyethyl acrylate   37, 60–61
–, MA-POSS   50–51
–, methyl methacrylate   24, 26, 37, 45–50
–, 2-[(perfluorononenyl)oxy]ethyl methacrylate   64
–, poly(dimethylsiloxane)   101
–, poly(sulfone)   79–80, 81
–, poly(vinylidene fluoride)   86
–, styrene   24, 35, 37, 38, 39, 55–56, 56, 65, 69, 70, 86, 97, 98
–, vinyl acetate   84, 85
n-Butyl acrylate homopolymerization   38, 39–40
t-Butyl acrylate
–, ABC triblock copolymers   56–58
–, copolymers with poly(ethylene glycol)   73–74
–, copolymers with styrene   56
n-Butyl methacrylate
– –, copolymers with methyl methacrylate   24, 65, 69
– –, copolymers with styrene   15, 17–18, 42, 51–52, 69
– –, end group elimination   42
– –, homopolymerization   42

ε-Caprolactone
–, as a counter radical   99

–, copolymers with 2,7-dibromo-9,9-
    dihexylfluorenes   149–150
–, copolymers with methyl
    methacrylate   148, 150–151
–, copolymers with styrene   99, 105, 148
–, graft copolymers   117, 149, 150–151
–, star copolymers   142
Carbosilane dendrimers   89
Chain topology   2
Chain transfer agents
–, in conventional free radical
    polymerizations   5
–, in controlled radical polymerizations   7,
    8
–, in block copolymerizations   85, 86
Condensation polymerization   80, 81, 82,
    149–150
Controlled radical polymerizations (CRP)
    6–12, 153
(see also ATRP, SFRP/NMP, RAFT, degenerative transfer)
– – –, atom transfer radical polymerization (ATRP)   10–11, 12, 13, 153
– – –, end group reactivity   28, 44–45
– – –, equilibrium in   6, 7, 28
– – –, reversible addition fragmentation
    chain transfer (RAFT)   11–12, 13,
    153
– – –, stable free radical (SFRP)/nitroxide
    mediated (NMP)   8–10, 12, 13, 153
– – –, termination in   6–7
Conventional free radical
    polymerizations   6
–, molecular weight control   5, 7
–, termination reactions   4
–, reactivity ratios   5
–, combination with CRP   83, 84–86
Copolymers
–, ABC triblock   56–58, 83–84, 101
–, acidic   69, 70–71, 72, 73, 74, 130, 141,
    144, 154, 156
–, alternating   14, 17, 19, 20, 25, 26, 27, 42,
    43
–, amino functional   32–33, 52, 146
–, amphiphilic   40, 42, 44, 56, 59–61, 63,
    67–68, 72–75, 90, 99–100, 118, 130, 140–
    141, 142, 144, 145, 148, 155
–, biocompatible   74, 75, 79
–, block, linear   1, 2, 5, 6, 14, 15, 16, 17, 19,
    20, 22, 24, 27, 30–106, 148–150,
–, fluorinated   44, 64–65, 86
–, gradient   1, 2, 5, 14–15, 19–23, 26, 27, 28,
    46, 48, 49–50, 60–61, 71, 120, 122, 137, 153,
    155, 156

–, graft   107–138, 147–148, 150–151
–, hybrid organic/inorganic   50–51, 75–76,
    83, 89, 100–101, 111, 133–137, 138
–, liquid crystalline   33–34, 37–38, 70
–, photochromic   15, 16
–, reactivity ratios   4–5, 14, 15, 20, 24, 25,
    118, 119, 122, 123,
–, star shaped   138–147
–, statistical   1, 2, 3,14–29, 35, 36, 65, 66, 71,
    82, 153, 154, 155
–, topology   2
–, water soluble   40, 59, 61, 63–64, 67, 70,
    74, 75
Coupling reactions   81, 82
β-Cyclodextran
–, as an ATRP initiator   75
–, copolymers with N,N-(dimethylamino)ethyl methacrylate   75
–, copolymers with methyl
    methacrylate   75
–, copolymers with poly(ethylene glycol)
    methacrylate   75
Cyclohexene oxide   95
– –, copolymers with styrene   95–96

Degenerative transfer   26, 69, 153
Dendrigraft copolymer   90, 113
Dendrimer   2
–, poly(ether)   87
–, carbosilane   88
DEPN   38, 43, 153, 167
Dicyclopentadiene   101–102
1,1-Dihydroperfluorooctyl acrylate   65, 130
1,1-Dihydroperfluorooctyl methacrylate
– –, homopolymerization   65
– –, copolymers with methyl
    methacrylate   65
– –, copolymers with N,N-(dimethylamino)ethyl methacrylate   65
N,N-Dimethylacrylamide
–, copolymers with methyl acrylate   62
–, copolymers with styrene   37, 41, 70
–, homopolymerization   38, 41, 61, 62
N,N-(Dimethylamino)ethyl methacrylate,
    copolymers with
– –, benzyl methacrylate   69, 70
– –, β-cyclodextrin   75
– –, 1,1-dihydroperfluorooctyl
    methacrylate   65, 68
– –, methyl methacrylate   59, 67
– –, styrene   59
N,N-(Dimethylamino)ethyl methacrylate
– –, copolymers with methyl
    methacrylate   59

Subject Index 187

– –, grafted from gold   132
– –, homopolymerization   42, 43

Electropolymerization   86–87
Emulsion   56, 136, 156
Ethyl acrylate
– –, copolymers with methyl acrylate   69
– –, homopolymerization   53
Ethylene glycol monomethacrylate monoperfluorooctanoate   64
4'-Ethylbiphenyl-4-(4-propenoyloxy-butyloxy)benzoate 37–38

Functionalization
–, carbosilane dendrimers   89
–, β-cyclodextrin   75
–, oligophenylenes   80, 82
–, poly(butadiene)   78, 96, 97
–, poly(dimethylsiloxane)   76, 100–101
–, poly(ester)   80
–, poly(ether) dendrons   87
–, poly(ethylene adipate)   74–75
–, poly(ethylene-co-butylene)   77
–, poly(ethylene glycol)   73, 74
–, poly(methylphenylsilylene)   81
–, poly(phenylenevinylene)   82
–, poly(propylene glycol)   74, 90
–, poly(sulfone)   79–80
–, poly(vinylidene fluoride)   86

Gradient copolymers
–, formation of   71
–, methyl methacrylate/n-butyl acrylate   46, 48, 49–50
–, methyl methacrylate/methacrylate-PDMS   27
–, methyl methacrylate/methyl acrylate   20
–, styrene/acrylonitrile   20–23, 97
–, styrene/methyl acrylate   20, 21, 22
Graft copolymers
– –, grafting from   108–116
– –, grafting onto   126–127
– –, grafting through   117–126
– –, ω-methacryloyl poly(caprolactone)/styrene   117
– –, ω-methacryloyl poly(dimethylsiloxane)/methyl methacrylate   27, 119–122
– –, ω-methacryloyl poly(ethylene)/n-butyl acrylate   124
– –, ω-methacryloyl poly(ethylene)/styrene   117
– –, ω-methacryloyl poly(ethylene glycol)/styrene   117
– –, ω-methacryloyl poly(ethylene oxide)/styrene   117
– –, ω-methacryloyl poly(isobutyl vinyl ether)   124, 126
– –, ω-methacryloyl poly(lactic acid)/methyl methacrylate   122–124, 125
– –, ω-methacryloyl poly(lactide)/styrene   117
– –, ω-methacryloyl poly(methyl methacrylate)/n-butyl acrylate   118–119
– –, ω-vinyl poly(styrene)/N-vinyl pyrrolidinone   118
– –, methods   107
– –, poly(2-(2-bromopropionyloxy)ethyl acrylate)/n-butyl acrylate   114–115
– –, poly(2-(2-bromopropionyloxy)ethyl acrylate)/p(n-butyl acrylate-co-styrene)   115
– –, poly(2-(2-bromopropionyloxy)ethyl acrylate)/styrene   114–115
– –, poly(dimethylsiloxane)/styrene   111
– –, poly(ethylene-co-glycidyl methacrylate)/methyl methacrylate   111
– –, poly(ethylene-co-glycidyl methacrylate)/styrene   111
– –, poly(isobutylene)/p-acetoxystyrene   115
– –, poly(isobutylene)/methyl methacrylate   110, 112
– –, poly(isobutylene)/styrene   110, 115
– –, poly(isobutylene-co-p-methylstyrene-co-p-bromomethylstyrene)/styrene   110
– –, poly(methyl methacrylate)/methyl methacrylate   114
– –, poly(propylene)/styrene   108–109
– –, poly(styrene-b-ethylene-co-propylene)/ethyl methacrylate   110
– –, poly(styrene)/n-butyl methacrylate   113
– –, poly(styrene)/methyl acrylate   110
– –, poly(styrene)/methyl methacrylate   110, 113
– –, poly(styrene)/poly(ether) dendron   126–127
– –, poly(styrene)/styrene   108, 110, 113, 114
– –, poly[(vinyl chloride)-co-(vinyl chloroacetate)]/n-butyl acrylate   109, 110
– –, poly[(vinyl chloride)-co-(vinyl chloroacetate)]/methyl acrylate   109
– –, poly[(vinyl chloride)-co-(vinyl chloroacetate)/methyl methacrylate   109

- -, poly[(vinyl chloride)-*co*-(vinyl chloroacetate)]/styrene 109
Grafting, from gold surfaces
-, *t*-butyl methacrylate 132
-, *N,N*-(dimethylamino)ethyl methacrylate 132
-, 2-hydroxyethyl methacrylate 132
-, isobornyl methacrylate 132
-, methyl methacrylate 132, 133
Grafting, from silicon surfaces
-, 1,1-dihydroperfluorooctyl acrylate 130
-, poly(acrylamide) 127–128
-, methyl acrylate 130
-, methyl methacrylate 128–129
-, methyl methacrylate/styrene 129
-, styrene 129, 130
-, styrene/*t*-butyl acrylate 130
-, styrene/methyl acrylate 130
-, styrene/methyl methacrylate 130–131, 132

Homopolymerizations
-, acrylonitrile 38
-, *N-tert*-butylacrylamide 61
-, *n*-butyl acrylate 38, 39–40
-, *n*-butyl methacrylate 42
-, 1,1-dihydroperfluorooctyl acrylate 65
-, 1,1-dihydroperfluorooctyl methacrylate 65
-, *N, N*-dimethylacrylamide 38, 41, 61, 62
-, *N,N*-(dimethylamino)ethyl methacrylate 42, 43
-, ethyl acrylate 53
-, *N*-(2-hydroxypropyl) methacrylamide 61
-, isoprene 39–40
-, 3-*O*-methacryloyl-1,2:5,6-di-*O*-isopropyl-D-glucofuranose 62–63
-, methyl methacrylate 42, 53, 55, 66
-, styrene 38, 53
-, vinyl acetate 84, 85
-, 4-vinyl pyridine 40, 60
Hydrogel 118
2-Hydroxyethyl acrylate copolymers 60–61
2-Hydroxyethyl methacrylate, copolymers with
- -, isoprene 40
- -, methyl methacrylate 25, 59–60, 148
- -, styrene 52
N-(2-Hydroxypropyl) methacrylamide 61
Hyperbranched polymers 89–91, 147

Iniferter 7
Isobornyl methacrylate 132
Isoprene, copolymers with
-, acrylic acid 40
-, *t*-butyl acrylate 40
-, 2-hydroxyethyl methacrylate 40
-, styrene 39–40, 99
Isoprene 39–40

Ligands, copper
-, 2,2'-bipyridine 10, 19, 24, 25, 26, 44, 51, 52, 53, 54, 55, 59, 61, 64, 67, 68, 73, 74, 77, 82, 85, 86, 87, 95, 98, 102, 103, 115, 127, 138, 142, 143, 144, 154, 166
-, 2,2'-di(5-alkyl)-2,2'-bipyridine 66, 68
-, 4,4'-diheptyl-2,2'-bipyridine 87, 102, 124, 128, 167
-, 4,4'-dimethyl-2,2'-bipyridine 54, 167
-, 4,4'-di(5-nonyl)-2,2'-bipyridine 24, 26, 47, 54, 85, 86, 91, 92, 94, 101, 119, 120, 123, 130, 138, 139, 167
-, 4,4'-di(tridecafluoro-1,1,2,2,3,3-hexahydrono)-2,2'bipyridine 65, 68, 167
-, *n*-octyl-2-pyridylmethyanimine 89, 168
-, *N,N,N',N'',N''*-pentamethyldiethylenetriamine 24, 26, 51, 54, 60, 67, 68, 73–74, 84, 85, 91, 110, 130, 131, 168
-, *N*-(*n*-pentyl) 2-pyridylmethyanimine 143
-, *n*-propyl-2-pyridylmethyanimine 75
-, 1,4,8,11-tetramethyl-1,4,8,11-tetraazacyclotetradecane 61, 62, 68, 168
-, tris [2-(dimethylamino)ethyl] amine 24, 26, 60, 61, 67, 133, 154, 168
-, tris(4,4'-dimethyl-2,2'-bipyridine) 53
Ligands, ruthenium
-, *p*-cymene 25, 26
-, trialkylphosphine 25
-, triphenylphosphine 25, 26, 44, 45, 54, 139
Liquid crystals 33–34, 37–38, 71
Living polymerization systems 6
- - -, anionic 4, 5, 6, 30, 76, 79, 96–99, 101, 106, 144, 147, 148, 153
- - -, cationic 4, 5, 6, 30, 72, 79, 91–95, 104, 106, 108, 115, 124, 130, 142, 147, 148, 153
- - -, equilibrium 6

Maleic anhydride 41, 77–78
Methacrylate-PDMS 27
Methacrylic acid 25
3-*O*-Methacryloyl-1,2:5,6-di-*O*-isopropyl-D-glucofuranose

Subject Index

–, homopolymerization 62–63
–, copolymers with styrene 63
Methyl acrylate, ABC triblock
  copolymers 56–57
Methyl acrylate, copolymers with
- –, dicyclopentadiene 101–102
- –, N,N-dimethylacrylamide 62
- –, ethyl acrylate 69
- –, ethylene glycol monomethacrylate mono-perfluorooctanoate 64
- –, methyl methacrylate 20, 47
- –, norbornene 101–102
- –, 2-[(perfluorononenyl)oxy]ethyl methacrylate 64
- –, poly(glycerols)
- –, poly(isobutylene) 92
- –, poly(tetrahydrofuran) 93–94
- –, poly(vinylidene fluoride) 86
- –, silicon 130
- –, styrene 19–20, 21, 22, 37, 55, 91–92, 97
Methyl methacrylate, copolymers with
- –, benzyl methacrylate 66
- –, butadiene 102–103
- –, n-butyl acrylate 24, 27, 45–50, 65
- –, n-butyl methacrylate 24, 65, 69
- –, ε-caprolactone 148, 150–151
- –, carbosilane dendrimers 89
- –, β-cyclodextrin 75
- –, 1,1-dihydroperfluorooctyl methacrylate 65
- –, N,N-(dimethylamino)ethyl methacrylate 59
- –, 2-hydroxyethyl methacrylate 25, 59–60, 148
- –, methacrylic acid 25
- –, ω-methacryloyl poly(dimethylsiloxane) 27
- –, methyl acrylate 20, 47
- –, poly(dimethylsiloxane) 101
- –, poly(isobutylene) 92
- –, poly(tetrahydrofuran) 93–94
- –, poly(vinylidene fluoride) 86
- –, pyrrole 86–87
- –, styrene 15, 17, 20, 25, 27, 37, 41, 42, 52, 53, 55, 69, 91–92, 97, 98
- –, 4-vinyl pyridine 60
Methyl methacrylate 42, 53, 55, 66
MOTEMPO 15, 19, 31, 36, 83, 168
Multifunctional initiators 147

Nanocomposites 133
Nanoparticles
–, grafting from 135–136

–, luminescent 135
Norbornene 101–102

Oligo(ethylene oxide) methacrylate
- –, copolymers with sodium 4-vinyl benzoate 64
Oligophenylenes
–, copolymers with styrene 80, 82
–, functionalization of 80, 82
Organic-inorganic hybrids 50–51, 75–76, 83, 100–101
OTEMPO 37, 43, 93, 99, 168

2-[(Perfluorononenyl)oxy]ethyl methacrylate 64
Persistent radical effect (PRE) 7
Poly(acrylamide) 127–128
Poly(2-(2-bromopropionyloxy)ethyl acrylate) graft copolymers 114–115
Poly(t-butyl methacrylate) from gold 132
Poly(dimethylsiloxane)
–, copolymers with n-butyl acrylate 101
–, copolymers with methyl methacrylate 101
–, copolymers with styrene 76, 83
–, functionalization of 76, 83
Poly(ethylene) graft copolymers 117, 124
Poly(ethylene adipate)
–, copolymers with styrene 75
–, transformation to a NMP counter radical 74–75
Poly(ethylene-co-butylene)
–, copolymers with styrene/maleic anhydride 77–78
–, functionalization with an ATRP initiator 77
–, functionalization with a RAFT initiator 77
Poly(ethylene glycol), copolymers with
–, benzyl methacrylate 74
–, t-butyl acrylate 73–74
–, β-cyclodextrin 75
–, methacrylic acid, sodium salt 74
–, styrene 73, 74
Poly(ethylene glycol), functionalization with
–, an ATRP initiator 73, 90
–, a NMP initiator 73, 99–100
–, a RAFT initiator 74
Poly(ethylene glycol) graft copolymers 117
Poly(ethylene oxide) graft copolymers 117
Poly(glycerols) 91

Poly((2-hydroxyethyl) methacrylate from gold 132
Poly(isobornyl methacrylate) from gold 132
Poly(isobutylene) 92–93
Poly(isobutylene) graft copolymers 110, 112, 115
Poly(isobutylene-*co-p*-methylstyrene-*co-p*-bromomethylstyrene) 110
Poly(isobutyl vinyl ether) graft copolymers 124, 126
Poly(lactic acid) graft copolymers 117, 122–124, 125
Poly(methyl methacrylate) graft copolymers 114, 118–119, 128–129, 132, 133
Poly(methylphenylsilylene)
–, copolymers with styrene 82
–, functionalization of 81
Poly(phenylenevinylene)
–, copolymers with styrene/chloromethyl styrene 82–83
–, functionalization of 82
Poly(propylene) graft copolymers 108–109
Poly(propylene oxide)
–, copolymers with styrene 74
–, functionalization with an ATRP initiator 74
Poly(styrene), graft copolymers with
–, *n*-butyl methacrylate 113
–, methyl acrylate 110
–, methyl .methacrylate 110, 113
–, poly(ether) dendron 126–127
–, silicon 129, 130–131, 132
–, styrene 108, 110, 113, 114
–, *N*-vinyl pyrrolidinone 118
Poly(sulfone) block copolymers 79–80, 81
Poly(tetrahydrofuran)
–, as an ATRP macroinitiator 93–94, 95
–, as an NMP counter radical 93
–, as a unimolecular initiator 93
–, copolymers with methyl acrylate 93–94
–, copolymers with methyl methacrylate 93–94
–, copolymers with styrene 93–94, 95
Poly(vinylidene fluoride)
–, as an ATRP initiator 86
–, copolymers with *n*-butyl acrylate 86
–, copolymers with methyl acrylate 86
–, copolymers with methyl methacrylate 86
–, copolymers with styrene 86
Pyrrole 86–87

RAFT 11–12, 13
–, block copolymers via 69–70, 71–72, 74, 77–78
–, graft copolymers via 120, 122
–, star polymers via 142
–, statistical copolymers 27
–, transfer constants 69
Ring opening metathesis polymerization 101–102, 150
Ring opening polymerization (ROP) 6, 30, 75, 79, 91, 93–96, 99–101, 105, 106, 142, 144, 147, 148, 149, 150, 151–152, 153, 155

SFRP/NMP 8–10, 12, 13
–, statistical copolymers 15–19, 28
–, block copolymers via 30–44, 70–71, 73, 74–75, 82–83, 88, 93, 95, 96–97, 99
–, graft copolymers 108–109, 114, 117–118, 126–127
–, star polymers via 138
Silicon 127–132
Star polymers
– –, amphiphilic 140–141
– –, arm first method 146
– –, from calixarene cores 139–140
– –, from dendritic cores 141
– –, from hyperbranched polymers 147
– –, from isobutylene cores 143–144
– –, from poly(ethylene oxide) cores 144–145
– –, hybrid organic/inorganic 138–139
– –, styrene 138
– –, via transformations 142
Styrene, copolymers with
–, *N*-(2-acetoxyethyl) maleimide 25, 27
–, *p*-acetoxymethylstyrene 24
–, *p*-acetoxystyrene 15, 24
–, acrylic acid 38
–, acrylonitrile 15, 16, 22–24, 34–35, 38
–, anionically prepared styrene 97, 98
–, 2,5-bis [(4-methoxyphenyl)oxycarbonyl] styrene 33–34
–, *p*-bromostyrene 15, 31
–, butadiene 39–40, 78, 96, 97, 102
–, *p-tert*-butoxystyrene 15, 32
–, *n*-butyl acrylate 25, 35, 37, 38, 39, 55–56, 56, 65, 69, 70, 86, 98
–, *t*-butyl acrylate 56
–, *N*-butylmaleimide 17
–, *n*-butyl methacrylate 15, 17–18, 42, 51–52
–, *p-tert*-butylstyrene 31–32
–, ε-caprolactone 99, 148

Subject Index

–, carbocationically prepared styrene 91
–, *p*-chloromethyl styrene 15, 30, 31, 40
–, chlorostyrene 31
–, cyclohexene oxide 95–96
–, *N*-cyclohexylmaleimide 16, 34
–, dicyclopentadiene 101–102
–, 2,7-dibromo-9,9-dihexylfluorenes 149–150
–, *N,N*-dimethylacrylamide 38, 41, 70
–, *N,N*-(dimethylamino)ethyl methacrylate 59
–, 1,4-dioxepane 95
–, epoxystyrene 18, 24, 25
–, 4'-ethylbiphenyl-4-(4-propenoyloxybutyloxy)benzoate 37–38
–, ethylene glycol monomethacylate monoperfluorooctanoate 64
– glycidyl acrylate 38
–, 2-hydroxyethyl acrylate 38
–, 2-hydroxyethyl methacrylate 52
–, isoprene 39–40, 99
–, maleic anhydride 17, 42, 77–78
–, 3-*O*-methacryloyl-1,2:5,6-di-*O*-isopropyl-D-glucofuranose 63
–, ω-methacryloyl poly(caprolactone) 117
–, ω-methacryloyl poly(ethylene glycol) 117
–, ω-methacryloyl poly(ethylene) 117
–, ω-methacryloyl poly(ethylene oxide) 117
–, ω-methacryloyl poly(lactide) 117
–, *p*-methoxymethylstyrene 24
–, *p*-methoxystyrene 15
–, methyl acrylate 19–20, 21, 22, 37, 55, 56, 91–92
–, methyl methacrylate 15, 17, 19–20, 25, 27, 38, 42, 52, 53, 55, 91–92, 98
–, 2-methyloxazoline 94–95
–, *p*-nitrophenyl methacrylate 52
–, norbornene 101–102
–, oligophenylenes 80, 82
–, oxazoline 148
–, 2-[(perfluorononenyl)oxy]ethyl methacrylate 64
–, *N*-phenylmaleimide 25, 27
–, phenyl oxazoline 148
–, phthalimide methylstyrene 32
–, poly(dimethylsiloxane) 76, 83
–, poly(ester) 80
–, poly(ether) dendrons 87–88
–, poly(ethylene adipate) 75
–, poly(ethylene-*co*-butylene) 77
–, poly(ethylene glycol) 73, 94

–, poly(isobutylene) 92
–, poly(methylphenylsilylene) 82
–, poly(phenylenevinylene) 82–83
–, poly(propylene oxide) 74
–, poly(sulfone) 79–80
–, poly(tetrahydrofuran) 93–94, 95
–, poly(vinylidene fluoride) 86
–, trimethylsilylmethyl 4-vinylbenzoate 18
–, trimethylsilylstyrene 24
–, vinyl acetate 84, 85–86
–, 4-vinyl pyridine 15, 40
Styrene, graft copolymers with
–, poly(2-(2-bromopropionyloxy)ethyl acrylate) 114–115
–, poly(dimethylsiloxane) 111
–, poly(ethylene-*co*-glycidyl methacrylate) 111
–, poly(isobutylene-*co*-*p*-methylstyrene-*co*-*p*-bromomethylstyrene) 110
–, poly(propylene) 108–109
–, poly(styrene) 108, 110, 113, 114
–, *N*-vinyl carbazole 17, 34
–, poly[(vinyl chloride)-*co*-(vinyl chloroacetate)] 109
Styrene 38, 53
4-Styrene sulfonate, copolymers with
–, 4-(dimethylamino)methylstyrene 35
–, sodium 4-styrenecarboxylate 35
–, vinyl naphthalene 35
Supercritical CO_2 66

TEMPO 8–9, 15–19, 24, 27, 28, 30–31, 32, 33–34, 35, 36, 37–38, 39–40, 41, 42, 43, 60, 61, 64, 70–71, 73, 74–75, 82, 87, 88, 93, 95, 96–97, 99, 108–109, 113, 114, 117, 118, 126, 129, 138, 147, 148, 153, 169
TMPAH 38, 39, 40, 42, 43, 69, 153, 154, 169

Vinyl acetate
–, copolymers with *n*-butyl acrylate 84
–, copolymers with styrene 84, 85–86
–, homopolymerization 84, 85
Vinyl benzyl chloride 89, 90, 91
–, copolymers with *N*-cyclohexylmaleimide 91
4-Vinyl pyridine
–, copolymers with methyl methacrylate 60
–, copolymers with styrene 40
–, homopolymerization 40, 60

Water borne CRP reactions 24, 35, 64, 65, 66, 68, 70

You are one **click** *away from a* **world of chemistry** *information!*

Come and visit Springer's
Chemistry Online Library

Books
- Search the Springer website catalogue
- Subscribe to our free alerting service for new books
- Look through the book series profiles

You want to order?   Email to: orders@springer.de

Journals
- Get abstracts, ToC´s free of charge to everyone
- Use our powerful search engine LINK Search
- Subscribe to our free alerting service LINK *Alert*
- Read full-text articles (available only to subscribers of the paper version of a journal)

You want to subscribe?   Email to: subscriptions@springer.de

Electronic Media
- Get more information on our software and CD-ROMs

You have a question on an electronic product?   Email to: helpdesk-em@springer.de

● Bookmark now:

http://www.springer.de/chem/

Springer · Customer Service
Haberstr. 7 · D-69126 Heidelberg, Germany
Tel: +49 6221 345-217/218 · Fax: +49 6221 345-229
d&p · 006756_001x_1c

Springer

Printing (Computer to Film): Saladruck Berlin
Binding: Stürtz AG, Würzburg

十